CELESTIAL HARVEST

300-Plus Showpieces of the Heavens for Telescope Viewing and Contemplation

James Mullaney

DOVER PUBLICATIONS, INC.
Mineola, New York

Bibliographical Note

This Dover edition, first published in 2002, is an unabridged, revised republication of the work originally published by the author in 1998.

Library of Congress Cataloging-in-Publication Data

Mullaney, James.
    Celestial harvest : 300-plus showpieces of the heavens for telescope viewing and contemplation / James Mullaney.
        p. cm.
    Originally published: Exton, Pa. : James Mullaney, 1998.
    Includes index.
    ISBN 0-486-42554-1 (pbk.)
    1. Astronomy—Observers' manuals. I. Title.

QB64 .M88 2002
522—dc21

2002067323

Manufactured in the United States of America
Dover Publications, Inc., 31 East 2nd Street, Mineola, N.Y. 11501

DEDICATION

* * * * *

To my loving wife, **Sharon McDonald Mullaney**, for her unfailing support and belief in the ultimate value of this labor-of-love to the worldwide fraternity of stargazers.

To my parents, **James and Ernestine Mullaney**, and my aunt and uncle, **Elva and Walter Jones**, for nurturing my interest in the stars from an early age - including the building of my first telescope.

To my sister, **Elvira Mullaney Schmitt**, and my brother, **John Mullaney**, for putting up with my nocturnal stargazing antics throughout our growing-up and school years.

To my uncle, **Robert Franks**, for giving me my first look at the Moon and Jupiter through his telescope, thereby starting me on my lifelong wondrous adventure as a "star pilgrim."

To **Arthur Draper**, past Director of the Buhl Planetarium, and **Nicholas Wagman**, past Director of the Allegheny Observatory, for early inspiration and the use of their marvelous facilities.

ABOUT THE AUTHOR

* * * * *

James Mullaney is an astronomy writer, lecturer and consultant who has published more than 500 articles and three books on observing the wonders of the heavens and logged over 20,000 hours of stargazing time with the unaided eye, binoculars and telescopes. Formerly Curator of the Buhl Planetarium & Institute of Popular Science in Pittsburgh and more recently Director of the DuPont Planetarium, he served as staff astronomer at the University of Pittsburgh's Allegheny Observatory, and as an editor for **Sky & Telescope, Astronomy** and **Star & Sky** magazines. One of the contributors to Carl Sagan's award-winning **Cosmos** PBS-Television series, his work has received recognition from such notables as Sir Arthur Clarke, Ray Bradbury, Johnny Carson, Dr. Wernher von Braun, and former student - NASA scientist/astronaut Dr. Jay Apt. His 40-year mission as a "celestial evangelist" has been to *Celebrate the Universe!* - to get others to look up at the majesty of the night sky and personally experience the joys of stargazing.

CELESTIAL HARVEST

300-PLUS SHOWPIECES OF THE HEAVENS FOR TELESCOPE VIEWING & CONTEMPLATION

by

James Mullaney

* * * * *

INTRODUCTION

The following compilation presents over 300 of the finest celestial wonders for viewing with common "backyard" telescopes (refractors, reflectors and catadioptrics in the 2- to 14-inch aperture range). Representing the culmination of the author's 40-year visual survey of the sky N of -45 degrees declination, it covers that 3/4ths of the entire heavens visible from mid-northern latitudes. This survey has resulted in the publication of more than 500 articles on "stargazing" over the years in such magazines as *Sky & Telescope*, *Astronomy, Star & Sky, Modern Astronomy, Review of Popular Astronomy, The Planetarian* and *Science Digest* - as well as the Sky Publishing reprint booklet *The Finest Deep-Sky Objects*, which has gone through three printings and been in use by amateur and professional astronomers around the world since 1966.

More than 3,000 deep-sky objects were examined with dozens of telescopes, the visual appeal of each being rated at the eyepiece on an excellent (E), good (G), fair (F), poor (P) basis. The E and G entries comprise this roster. All objects were observed when on or near the meridian (and so at their highest in the sky), and under conditions of average or better seeing/atmospheric steadiness (mainly doubles and tight clusters) and/or transparency/clarity (mainly clusters, nebulae and galaxies). The author's various observing sites were nearly all within the suburbs of large cities in the northeastern United States and, therefore, under conditions of moderate to heavy light pollution. This compilation begins with six showpieces of the solar system, followed by 336 deep-sky wonders - those objects lying beyond the confines of our Sun's family. (There actually are 370 in all, since a number of the entries have two or more related objects in the same eyepiece field, such as the Andromeda Galaxy and its companions.) The latter are arranged alphabetically by constellation and consist of the following nine classes or types: first-magnitude/highly-tinted and/or variable (single) stars, double and multiple stars, asterisms and stellar associations, open star clusters, globular star clusters, diffuse nebulae, planetary nebulae, supernovae remnants, and galaxies. Double and multiple stars dominate this compilation due to their great profusion in the sky (its tinted jewels!), as well as their endless variety and easy visibility on all but the worst of nights.

An abbreviated descriptive format is used to accommodate as much observing information as possible in a manner convenient for both field and armchair use (i.e. reading prior to going to the telescope). **Those objects (half of the total) indicated by an exclamation mark (!) represent the "best of the best" for showing visitors at your telescope, as well as at astronomy club star parties and observatory/planetarium public viewing nights**. In addition to the 300-plus showpieces themselves, nearly two dozen other special objects are listed that - while in most instances not visually impressive at the eyepiece due to large apparent size, faintness, etc. - are still noteworthy for the various reasons given under "Remarks" and should be included in any celestial roundup like this one for completeness. They are indicated by an asterisk (*) before their descriptions.

To locate these wonders, this compilation should be used in conjunction with a good star atlas or map such as *Norton's 2000.0, The Cambridge Star Atlas, Sky Atlas 2000.0* or *Deep Map 600* (available from many telescope dealers and through bookstores) to "star hop" your way from bright naked-eye stars to the object you seek. Note that many of these showpieces are directly visible to the unaided eye on a dark (i.e. without bright moonlight) clear night, and are wonderful sights in binoculars and low-power (X), wide-field (WF) or richest-field (RFT) telescopes. Most of the objects presented here can be seen in a small glass, many of them even under bright-sky conditions. Should your instrument have setting circles (standard or digital) or a computerized star catalog connected to it, you can "dial up" these targets directly without recourse to a map or atlas (which - to many of us purists - takes much of the fun and adventure out of celestial exploration!).

As the famed observer William H. Pickering pointed out in his classic 1917 list of *The Sixty Finest Objects in the Sky*: "Incidentally, it may be mentioned that for the casual visitor who merely wishes to see something brilliant and surprising, few objects excite more admiration than a first-magnitude star thrown slightly out of focus." This is true especially when the star is near the horizon while rising or setting, at which time glorious prismatic rays flash and dance from its flaming heart!

A few final words: Enjoy these wonders for yourself, but also share them with others - and together celebrate the universe! Look at them with *both* your mind and your sight, contemplating the significance of what you're seeing. Realize that you are a part of them (you're made of "star stuff"), and are in *direct physical contact* with these objects via the amazing "photon connection" - the wave-particles hitting the retina of your eye were once inside of them! And finally, think about Who created all of these wonders and Who keeps them all shining! Clear skies and many happy "adventures in starland" as you explore the awesome "nocturnal wonderland" of the heavens! (See Page 97)

SOLAR SYSTEM WONDERS

* * * * *

OBJECT	DESCRIPTION

Sun   Our Daytime Star! Sunspots, granulation, faculae, solar rotation, flares, limb darkening (creating a 3-Dimensional effect!), prominences (with H-Alpha filter), transits of the inner planets (and other unidentified bodies - more than 600 of them having been reported in the literature over the past three centuries!), both partial and total solar eclipses, etc. **Extreme caution must be exercised - using proper filter and/or projection techniques - or instant blindness may result!** Our average distance from this radiant "lifestar" is 93,000,000 miles (or 8 light-minutes!).

Moon   Earth's Satellite. The most-observed and fascinating of all heavenly bodies! Ever-changing phases, lava plains ("seas"), craters, mountains, valleys, domes, pits, rilles, transient lunar phenomena (or TLPs - flashes, obscurations, colorations, moving lights, etc.), librations, conjunctions and occultations (of planets, stars and deep-sky objects - especially star clusters), both partial and total lunar eclipses, etc. With a good lunar map, embark on a nightly "sightseeing tour" of this fascinating alien landscape! Average distance from us 239,000 miles - making it the closest of all celestial bodies (with the exception of "Earth-grazing" asteroids!).

Venus   The Morning/Evening "Star" - and third brightest object in the heavens after the Sun and Moon. Changing phases, terminator irregularities, cloud shadings, the "spoke system," the "ashen light," the phase or "dichotomy" anomaly, cusp extensions, etc. Minimum distance from us 25,000,000 miles (much closer than Mars is - see below), at which time it displays a magnificent crescent that's visible in binoculars and has even been seen with the unaided eye!

Mars   The Red Planet (actually orange!). Largely disappointing except around times of opposition when closest to us (minimum distance 34,000,000 miles). Then, polar ice caps, deserts, greenish-blue and brown markings, streaky "canals," dust storms, clouds, the "blue clearing" of the atmosphere, rotation of planet, springtime "wave of darkening," etc. are visible. Mysterious flashes have been reported on both the surface and limb of the planet. (Its satellites Phobos and Demos can be glimpsed in 10-inch or larger instruments by a trained observer under good conditions - definitely not sights to impress visitors with!)

Jupiter   The Giant Planet. Most "active" and fascinating of all the planets! Four bright Galilean satellites Io, Europa, Ganymede and Callisto (all visible in binoculars) do a nightly dance around the planet and undergo eclipses, transits, occultations, cast their shadows on cloudtops of planet, and (every six years) mutual phenomena of each other! In good seeing at high power, they show tiny (non-stellar) disks. At least one harbors an ice-covered water ocean! The Great Red Spot, colorful cloud belts, bands, zones, festoons, limb and polar darkening, rotation of planet, central meridian transits, polar flattening, etc. Minimum distance from us 365,000,000 miles.

Saturn   The Ringed Planet. The most beautiful and stunning sight in the entire heavens! Multiple, razor-thin ice-rings with their Cassini and other less-obvious divisions, flattened globe with limb and polar darkening, shadows of ball on rings and rings on ball, retinue of at least five satellites visible (of the 16 now known!) including cloud-shrouded Titan, rings edge-on every 15 years with moons threading ring like beads on a string, etc. Minimum distance from us 750,000,000 miles.

The changing phases of **Mercury**'s pale-pink little disk, **Uranus**'s greenish and **Neptune**'s bluish minute disks, remote star-like **Pluto**, the brighter asteroids such as **Ceres**, **Pallas**, **Juno** and **Vesta**, and the occasional (usually unpredictable) appearance of a bright comet are other solar system targets of interest. However, these objects are not real showpieces in the sense used here (unless the comet happens to reach naked-eye brightness, at which time it may become a wondrous sight in binoculars and telescopes!).

DEEP-SKY WONDERS

* * * * *

The standard IAU (International Astronomical Union) three-letter abbreviations are used for constellation names.* For each object listed: its position in Right Ascension (RA - hours & minutes) and Declination (DEC - degrees & minutes) is given for Epoch 2000.0;** apparent visual magnitude/s; spectral type/s for first-magnitude, red/variable and double stars; angular size or separation in minutes (') or seconds ('') of arc; and official or popular name/s (if any). Primary data sources used were: *Sky Catalogue 2000.0* and *Burnham's Celestial Handbook*. Quotes from noted observers past and present (the Herschels, Lassell, Lord Rosse, Smyth, Chambers, Webb, Struve, Pickering, Barns, Olcott, Flammarion, Copeland, Bernhard/Bennett/Rice, Houston, Lorenzin, Mallas, Harrington, Eicher, O'Meara, etc.) are included under "Remarks." Approximate distance (D) in lightyears (LY) is also given for many entries. Messier (M), New General Catalog (NGC), Index Catalog (IC), Melotte (MEL) and Collinder (COL) designations are used for clusters, nebulae and galaxies. Checking one of the popular planispheres (a rotating star chart that is set to date and local time), or the monthly star maps in *Sky & Telescope* and *Astronomy* magazines, will show which constellations (60 of the 88 officially-recognized ones being represented here) are visible and well-placed in the sky on any given night for exploring their celestial treasures. Dotted lines under the object designation are for personal notes (date seen, telescope and magnification used, sky conditions, your own eyepiece impressions, etc.), while the # in ( ) indicates how many of 21 standard classic & modern deep-sky showpiece lists include that object - providing a quantitative index of its popularity (as does the length of the descriptive remarks for each entry!). **Note that those double and multiple stars designated by ! and having components 3" or less in separation will generally require at least a 6-inch aperture, good optics, relatively high magnification (150X or more) and a very steady night to be seen in their full glory as a finest showpiece!** (* See Page 94 for listing. ** For multiple-object entries, position given is for the primary- usually brightest - object in the field.)

KEY:			
SS=First-Magnitude/Highly-Tinted &/or Variable (Single) Star	OC=Open Star Cluster	PN=Planetary Nebula	EG=Elliptical Galaxy
DS=Double or Multiple Star	GC=Globular Star Cluster	SR=Supernova Remnant	IG=Irregular Galaxy
AS=Asterism or Stellar Association	DN=Diffuse Nebula	SG=Spiral Galaxy	MW=Milky Way

OBJECT/CONST	RA	DEC	TYPE	MAG/SPEC	SIZE/SEP	NAME/REMARKS
Gamma AND (20)	02 04	+42 20	DS	2.3, 5.5 K3II, B9V+A0V	10"	**! Almach** Magnificent topaz & aquamarine duo! "Orange & emerald." "Gorgeous colorings - burnished gold & cerulean blue. A notable doublet." "Clearly the finest colored double in the sky." "Beautiful pair! Gold/blue!" "Brilliant pair." "One of the most beautiful double stars in the heavens." "Beautiful contrast." "Contrasted colors of that lovely star - orange, emerald green; splendid...beautiful." "One of the most picturesque for small glasses." B is a close binary (companion mag. 6.3, separation ranging from 0.1" to 0.6") with a 61-year period. "B-C looks like a 'dumpish egg'" with 520X on an 8-inch when 0.4" apart. Can be split in a good 8- to 10-inch when around its widest separation, which last occurred in 1982. "Forms an admirable test object, but it is important to note the difference between one long disc crossed by a dark interference line, which is sometimes called division, & two round discs with black sky between them." This duo is now closing to another minimum in 2013 & is becoming tougher with each passing year. The three suns look "Orange, sea-green & blue" respectively. The A-B combo is fixed & moving through space together as a "common proper motion" (or CPM) pair. "Stunning appearance that rivals Albireo" (Beta CYG). Actually brighter, tighter & more intensely tinted than that famed pair, Almach ranks as the finest colored double for large instruments. "A beautiful DS. The brighter star is golden yellow or perhaps slightly orange, & the companion...a definite greenish-blue (the blended hues of B-C!). The color contrast is unusually fine

& often seems more striking with the eyepiece very slightly displaced from the position of sharpest focus" - as is the case for viewing stellar tints in general. The trio "Creates a wonderfully colored scene." The A-B components are separated by about 800 x the Earth-Sun distance, while B-C are 30 x it apart. Star B is itself a visually unresolved spectroscopic binary with rapid 2.7-day orbital period, so here we're actually beholding an amazingly complex four-star quadruple system! Superb even in 2-inch. D=260LY

M31/M32/M110 AND	00 43 +41 16	SG	3.5	178'X63'	
(21/12/11)		EG	8.2	8'x6'	
		EG	8.0	17'x10'	

! Andromeda Galaxy The Great Spiral of Andromeda & its elliptical companions! Largest, brightest & finest of all the galaxies (after the MW!). Nuclear bulge, spiral arms & dust lanes all readily visible! "One of the grandest in the heavens; long, oval, or irregularly triangular, ill-bounded, & brightening to the centre." "Queen of the nebulae." "Huge ellipse." "Overpowering" sight. The "Little Cloud" of the ancients. "Emits a radiant beam." "An object of increasing interest & wonder. Demands periodical observation under varying conditions to unfold the true glory of this celestial paragon." "An oval glow cast in an eerie greenish-white hue. Lovely spiral pattern." "The Great Nebula in Andromeda...grand spiral." "Impressive in a small telescope, although beginners may be disappointed they cannot see its details (as shown in its magnificent photos!)...the central condensation appears very intense & starlike." "One of the most beautiful deep-sky sights of all. The chief glory of amateurs." Very slowly sweeping the telescope back & forth across M31 actually *will* reveal much faint, subtle detail, including sharply-defined dust lanes between the arms (especially noticeable N of the central bulge). A "Fairly defined naked-eye spectacle" & binocular wonder! It has been traced to more than 5 x 1 degrees in overall size with the latter! The tiny, intense star-like nucleus is visible at very high X. A *6th-magnitude* (!) supernova (now designated S AND) appeared near this nuclear hub in 1885. "Small telescopes resolve M31's pronounced core as a bright 10' (diameter) sphere engulfed in the elongated glow of the galaxy's disk. A 4-inch telescope adds a distinctive dark lane girdling the plane of the galaxy to the NW of the core." Positioned about 24' SE of M31's nucleus is M32 - a small, bright & obvious egg-shaped glow. A "Beautiful galaxy. One of the best examples of an elliptical. Oval form & bears magnification well." "An oval patch...highlighted by a bright, nonstellar core." Much dimmer due to its larger apparent size & corresponding low surface brightness, M110 is located 36' NW of M31's hub. It's "Noticeably fainter than its more flamboyant neighbor (M32)... appears as a faint glow stretching NW-SE." "Large, faintish nebula of oval form - a mysterious apparition." "Pearl white." "Seems to sparkle." "Most recent addition to the Messier catalog (& hopefully the last) was proposed in 1967. An impressive sight in 4-inch." All three galaxies easily glimpsed in a 3-inch glass at 45X. Gazing upon M31 itself, you're seeing the combined radiance of some 300 *billion* stars! The diameter of this vast pinwheel is over 100,000 LY! Biggest & brightest member of our "Local Group" of galaxies, this "stain of magical dim light" serenely sails the ocean of intergalactic space tipped 15 degrees towards our line of sight. Near its S end, close to the W rim is M31's brightest starcloud NGC 206, visible in a 6-inch. D=2,400,000LY

NGC 752/56 AND (12/2)	01 58 +37 50	OC DS	5.7 5.7, 5.9 K0III, M0III	50' 190"	! Sprawling clan of over 60 stars with wide, matched orange DS "parked on SW edge." "Nebulous unit, curious groupings." "One of the finest large OCs in the sky." "Big interesting, underrated object - a real showpiece in an RFT." "A field-filling City of Stars - unusual, big, beautiful OC." "Unusually large...rich region." With 56 AND on its outskirts, "An imposing spectacle."! The two objects make a great combo, but are physically unrelated: D=360LY & 1,200LY, respectively! While here, also check out the neat bluish-white pair 59 AND (mags 6.1 & 6.8, 17" apart), NE of NGC 752 itself.
NGC 7662 AND (14)	23 26 +42 33	PN	8.5/13.2	32"x28"	! **Blue Snowball** Striking soft-blue planetary with star nucleus & internal structure. "One of Autumn's outstanding PNs." "Cobalt blue...flashing light." "Spectacular object...green." "Bright, bluish-green, elliptical...variable nucleus & two oval rings." "Remarkably bright, slightly elliptical." "Small but very bright...bluish disc with woolly border & suspicion of dark centre." Small blue dot in 4-inch, impressive sight in 8-inch & displays perforated disk (annular) with faint central star in 12-inch at high X. Its "Soft blue-green glow sets it apart from surrounding stars" even in the smallest of glasses. "Bluish & bright." "On photos, bears a fanciful resemblance to a lily." (And also as seen in big observatory-size instruments!) One of the brightest of its class & easily found above the huge 'Great Square' asterism of Pegasus. D=5,600LY
Beta AND/NGC 404 (2)	01 10 +35 37	SS/EG	2.1/11	3'x3'	* **Comet Komorowski** (& many others!) Bright star with dim EG in field 6' to NW. Sky's finest example of a 'false comet.' "Easy to locate but hard to see." "Good test of light-gathering ability of a telescope." Not plotted on most star atlases as its image overlaps that of Beta itself, raising false hopes of discovery upon seeing it! "Finding it was like discovering a diamond under a stone." Visible in 3-inch, obvious in 6-inch.
NGC 891 AND (5)	02 23 +42 42	SG	10.0?	14'X3'	Due to rich starry foreground creating a striking 3-D depth effect, the most picturesque classic edge-on spiral with dust lane in the entire sky. But "Not an easy object in small telescopes." "Sliver of pale light in 2- or 3-inch...6-inch...easily shows it." Tough in an 8-inch under light-polluted skies - needs at least a 10- to 12-inch & very dark night to appreciate. "One of the...legendary faint galaxies." "This wonderful object was indistinctly seen...on a glorious night" in a 6-inch. "A cigar-shaped sliver of dim grey light...a long, thin, almost ghostlike glow" in an 8-inch. The dark lane itself is very elusive visually. "Extraordinary object." "Beautiful...a favorite." D=13,000,000LY

(Please see Page 95 for ANT)

Zeta AQR (21)	22 29 -00 01	DS	4.4, 4.5 F1IV, F5IV	2"	! Matched tight greenish-white duo. "So splendid & close a pair." "Celebrated binary." "Beautiful." "Very fine object, easy with small aperture." "Twin white stars." Split in 3-inch at 90X on steady night - excellent test object. Set in starry naked-eye triangle. Orbital period 850 years, slowly opening to a maximum of 6" in 2265. D=76LY
41 AQR (8)	22 14 -21 04	DS	5.6, 7.1, 9.0 K0III, F2V	5", 212"	A-B components "Topaz yellow & sky blue." "Yellow-orange pair." "Reddish & blue." "Beautiful...topaz yellow, cerulean blue." "White, blue." A third fainter star "makes it a pretty group." All three components clearly seen in a 4-inch at 45X.

53 AQR (7)	22 27 -16 45	DS	6.4, 6.6 G1V, G2V	3"	Close, matched yellowish-white pair. "Neat DS, beautiful object, both pale white." Both stars "bright yellow." Just split in a 2.4-inch (60mm) refractor at 60X.
94 AQR (9)	23 19 -13 28	DS	5.3, 7.3 G5IV, K2V	13"	Lovely but neglected contrasting pair for all size scopes. "Colorful." "Good contrast." "Pale rose & light emerald or blue." "Yellow with perhaps reddish glare, greenish." "Yellow-white, blue." Nice sight in 2-inch glass at 25X - very attractive in 6-inch.
107 AQR (9)	23 46 -18 41	DS	5.7, 6.7 F0IV, F2III	7"	A "Pretty pair!" Tints given range from both white, to red & blue, to yellowish-white & lilac. Color changes suspected by early observers. "Very neat DS." "Fine colors."
M2 AQR (19)	21 34 -00 49	GC	6.5	13'	! "Magnificent ball of stars." "A glowing opalescent mass." "Like a heap of fine sand." "Beautiful." "Imagination cannot but picture the inconceivable brilliance of their visible heavens to its animated myriads." "Little ball of glowing mist in 2-inch ...in large telescopes...a wonderful sight." "Its magnificent stellar components resolvable in 9-inch or larger instrument & affording a superb view." Small glass shows a "granulated aspect, the precursor of resolution." "Not resolved in 4-inch." "One of those happy objects...a delight...spectacular." A starburst in 12-inch! D=37,000LY
NGC 7009 AQR (19)	21 04 -11 22	PN	8.3/11.5	25"x17"	! Saturn Nebula "Bright blue-green ellipse." "Prodigious! Elliptical, pale blue." "Magnificent." "Strikingly beautiful object in large scopes, shining with a vivid green fluorescent glow." "Bright to the very disc...pale blue." One early observer "saw it sparkle & thought it a heap of stars." "Bearing magnifying more like a planet than a common nebula. One of the finest specimens of these extraordinary bodies." Its ansae or "edge-on ring" extensions need at least an 8-inch & a dark steady night to glimpse. The featureless disk itself has an eerie radiance to it that's obvious in all size scopes but is especially striking in larger apertures. Pretty little blue egg in 3-inch at 45X - a truly awesome sight in 13-inch at 270X! A very "Distinguished PN"! D=3,000LY
NGC 7293 AQR (10)	22 30 -20 48	PN	6.5/13.5	16'x12'	* Helix/Sunflower Nebula Largest (apparent size half that of Moon!), brightest & closest (D=500LY) of the planetaries, but visually "a tricky customer" due to very low surface brightness. A huge "celestial smoke ring." "Sprawling." "Annular misty patch." This annular or ring-like appearance is not generally easy to make out visually regardless of what size instrument is used. Although the Helix has been seen in large binoculars (!), low-X /wide-field telescopes give the best view. Strangely missed by both the Herschels in their great visual surveys of the heavens (conducted with home-made metal-mirrored reflectors ranging from 6- to 48-inches in aperture!). "Unlike any PN in the sky - 1/4 degree across!" Needs very dark (Moonless) transparent night.
Alpha AQL (2)	19 51 +08 52	SS	0.77 A7V	---	Altair The "Flying Eagle Star." This "ice-blue gem" belongs to both the exclusive 'First-Magnitude Club' & to the huge 'Summer Triangle' asterism. "Brilliant stellar lighthouse." Host-star setting for sci-fi classic *Forbidden Planet*. Has very faint 10th-mag. optical companion at a distance of 152." Spins in just 6.5 *hours* - its equatorial

diameter is twice that of polar! "Central of three stars representing an eagle in flight - one of the few skiey similitudes that seems to be even remotely justified." "With its two attendant stars, one on either hand, presents a charming appearance." D=17LY

Object	Coord	Type	Mag	Sep	Description
15 AQL (4)	19 05 -04 02	DS	5.5, 7.2 K0, K1III	38"	Wide tinted duo. "Gems in a royal crown." "Yellow & ruddy purple or reddish lilac." "White & lilac." "Both yellow." Very easy in 2-inch glass. Optical (unrelated) pair.
57 AQL (9)	19 55 -08 14	DS	5.8, 6.5 B7V, B8V	36"	Another pretty, easy double for small scopes! Has "distinctly contrasted" but subtle tints. "A color-variant." "Colors variously recorded." "Colors curious." Primary usually described as pale yellow, while the companion is variously listed as bluish, greenish or lilac. Early observers were told: "Should be watched" for possible change. "So easy & wide an object - both blue." Attractive pair in 3-inch glass at 30X.
Struve 2404 AQL (6)	18 51 +10 59	DS	6.9, 8.1 K5, K3	4"	Relatively faint, tight combo with surprising hues. "Colors remarkable." "Azure & gold." "Colors marked, orange & reddish, close." "Orange pair." Striking in 8-inch.
V AQL (1)	19 04 -05 41	SS	6.6-8.4 N7.7 (C5)	---	A lovely glowing ember! "Intense fire red." "Deep red color." One of reddest stars in *Sky Cat 2000.0*. "Fine spectrum." Variable, like most rubies. Beautiful in 6-inch.
NGC 6709 AQL (4)	18 52 +10 21	OC	6.7	13'	Aquila's only good - but often ignored - stellar clan. Contains some 40 stars set in a "Beautiful wide group." "Kinda out there by its lonesome." "Scattered star cluster in fine field." "Alluring expanse follows." Nice in 10-inch at 50X. D=3,100LY
Gamma ARI (21)	01 54 +19 18	DS	4.8, 4.8 B9V, A1	8"	! **Mesarthim** Stunning, perfectly-matched silvery duo. "One of the finest pairs in the sky." "Suggests two little girls in blue." "The Cat's Eyes." "Certainly a beautiful pair." "Famous." "Striking twin whites." "Identical pearly twins." One of the first DSs discovered (1664). Lovely even in a 2-inch at 25X & spectacular in all apertures! Separation slowly closing; gravitationally bound together as a CPM pair. D=160LY
Lambda ARI (10)	01 58 +23 36	DS	4.9, 7.7 F0V, G0	37"	Wide, easy pair. "Striking color contrast." "White, olive or azure." "Yellow, & bluish, greenish or reddish-lilac." Good in 2.4-inch at 25X. CPM duo. D=105LY
30 ARI (9)	02 37 +24 39	DS	6.6, 7.4 F5V, F6III	39"	Another easy, wide tinted combo. "Colorful." Topaz-yellow & pale grey, yellow & pale lilac, yellow-white & white or azure are among the many different hues seen here.
Alpha AUR (2)	05 17 +46 00	SS	0.08 G6III+G2III	---	! **Capella** The "She Goat Star." Also the "Shepherd's Star." Radiant golden-yellow orb! "Very Brilliant." "Yellow blaze. Brightest G-type star in N sky." "Auriga's brightest jewel." An unresolved (spectroscopic) binary - dual solar-type stars in close 104-day orbit with 90 & 70 x Sun's brightness - 13 & 7 x its size! Ptolemy & other ancient astronomers described Capella as "red"! In more recent times it's been called simply "white"! "Described as red...it afterwards became yellow, & is now (1886) a pale blue."!? But most see it as the lovely topaz stellar gem that it truly is. D=42LY

Name	Coord	Type	Mag	Sep	Description
Theta AUR (9)	06 00 +37 13	DS	2.6, 7.1 A0II, G2V	4"	! Radiant, tight magnitude-contrast pair for steady nights. "Tough" - typically requires at least 6-inch. "Neat DS - brilliant lilac, pale yellow." "Blue & yellow." There's also 10th- & 11th-mag. stars at 130" & 50" distance, respectively. A very beautiful sight in 12-inch & larger apertures at 100X & up - given good seeing! CPM duo. D=110LY
14 AUR (9)	05 15 +32 41	DS	5.1, 7.4-7.9 A9IV, A2	15"	Neat pair. "Pale yellow, lilac" or "orange." Also "Greenish, blue-white." B variable. Also an 11th-mag. companion at 13" distance which has "a peculiar deep purple tint."
Struve 644 AUR (2)	05 10 +37 18	DS	6.7, 7.0 B2II, K3	1.6"	Lovely tight, color-contrast pair. "Gold, bluish-red; remarkable & constant colors." "Beautiful DS - topaz & amethyst." Best in 10-inch & larger apertures. Little-known & seldom observed - just one of the sky's many overlooked tinted stellar jewels!
UU AUR (5)	06 36 +38 27	SS	5.3-6.5 N3 (C5)	---	! Red "carbon" star - a celestial stoplight! "Beautiful red color." "Bright orange." Tint obvious in binoculars & a fine sight in even the smallest of scopes. Depth of color seems to vary with atmospheric conditions, changes in star's brightness & size of glass used - as is often the case for members of this class. Don't miss this gem!
M36 AUR (13)	05 36 +34 08	OC	6.0	12'	! "Beautiful assemblage of stars...very regularly arranged." "Impressive collection of 60 stars." "Not overly impressive." "A splendid cluster in a rich though open splash of stars." "Superb group." "Offers a rich harvest of double stars" - one of which is the 8th-mag. pair Struve 737 (separation 11"). "A gentle stellar fog. On exceptionally clear nights takes on an almost 3-D effect." "A grand view at low X, with marvelous color contrasts among the stars. Outward streamers of faint stars give a crab-like appearance." Considered the least striking of Auriga's well-known trio of Messier OCs, all of which are visible in binoculars. Lovely in 3-inch at 30X. D=4,000LY
M37 AUR (20)	05 52 +32 33	OC	5.6	24'	! Rich & uniform swarm - best of AUR trio. "Superb." "A magnificent object, the whole field being strewed, as it were, with sparkling gold-dust...resolvable into about 500 stars. Even in smaller instruments extremely beautiful, one of the finest of its class. Gaze at it well & long." "Wonderful loops & curved lines of stars." "A joy to behold." "A diamond sunburst!" "Very beautiful!" "Resolves into infinitely minute points of lucid light." "Puts on a dazzling show." At least 150 suns visible in small scopes, with red-giant star near center (as is often the case) "like a ruby on a field of diamonds." "A central 9th-mag. stellar ember, beams a radiant orange-red." "One of the finest OCs in the heavens...has a very elliptical outline with bunches of bright stars at ends." Like rich M11 in SCT (the Wild Duck Cluster), M37 is thought to be intermediate between the OCs & GCs by some astronomers. Superb in 8-inch. D=4,500LY
M38 AUR (17)	05 29 +35 50	OC	6.4	21'	! A "Noble cluster arranged as oblique cross. Larger stars dot it prettily with open doubles. Glorious neighbourhood." "Commanding." "Bright stars ranged against stardust." "Unusual shape." "Fine grouping of stellar fireflies." "A beautiful cluster in a splendid field...square-shaped with clump of stars at each corner." "Magnificent...

over 100 softly blazing stars evenly compressed into a glowing ball." Great in 5-inch at 50X. The tiny OC NGC 1907 (8th-mag., 7' dia.) located 0.5 degree S. D=4,000LY

Name	Coords	Type	Mag / Spectral	Sep	Description
Alpha BOO (2)	14 16 +19 11	SS	-0.04 / K2III	---	! **Arcturus** The "Bear Keeper Star." "Dazzling, distinctive orange hue unmistakable." "Splendid gem." "A noble object at all times." Other colors given are reddish-yellow, golden-yellow, topaz, peach, rosy champagne & that of a HP sodium street lamp! An orange giant, 25 x size of our Sun. High space velocity - over 2" per year, or Moon's apparent diameter in 900 years! Fourth brightest star in the heavens. D=37LY
Epsilon BOO (19)	14 45 +27 04	DS	2.5, 4.9 / K0II, A2V	2.8"	! **Izar** Magnificently-tinted but tight pair. "Most beautiful yellow, superb blue." "Light yellow, greenish." "Orange & green - a superb object." "Gold & blue test star." "Striking pair - pale orange, blue-green." "Lovely object. Pale orange, sea green; the colours being distinct & strongly contrasted." "Celebrated DS." Called "Pulcherrima" or "the most beautiful one" by Struve. "Fine contrast of yellow & green." "Bright yellow & marine blue." Rather difficult in apertures under 4-inches. A 3-inch at 150X shows two beautifully-colored diffraction disks nearly in contact! Extremely slow-orbiting physical pair with several-thousand-year period. D=160LY
Xi BOO (18)	14 51 +19 06	DS	4.7, 7.0 / G8V, K4V	6"	! Beautiful nearby binary system, with 150-year period. The "Colors are breathtaking. Strong yellow, vividly reddish-orange...unforgettable." "Orange & purple." "Yellow, reddish purple." "Showpiece double for small scopes - yellow & orange." "Impressive yellow & violet couple." "You'll be enchanted by the colors...secondary glows with an eerie reddish-violet light." Separation varies from 7" (last in 1977) to 2" (next in 2064). Currently easy in 3.5-inch at 40X, exquisite in 6-inch at 50X. D=22LY
Mu BOO (13)	15 24 +37 23	DS	4.3, 7.0, 7.6 / F0V, K0, K0	108", 2"	! Neat triple star with B-C close pair in 260-year orbit resolved in 3-inch at 100X. "All one vast system." "Attractive." "Pre-eminent quadruple (??) system." Colors here are yellow-white, orange & orange - in perfect agreement with spectral types! "Needs at least 100X for notched split." Actually, 75X does it in good seeing. D=95LY
Kappa BOO (8)	14 14 +51 47	DS	4.6, 6.6 / A8IV, F1V	13"	! Easy pair for 2-inch scope with hard-to-pin-down subtle tints. Both off-white, with greenish & bluish, or pale-yellow & bluish usually seen. "Yellow-white & purple" also given. "Beautiful object." "Very fine." Striking in 4-inch & larger apertures.
Pi BOO (15)	14 41 +16 25	DS	4.9, 5.8 / B9, A6V	6"	! Beautiful even in a 2.4-inch glass - closer version of Kappa BOO. Bluish white, pale orange or "tawny or ruddy." "Both white." "Pleasing close double." "Bright pair."
Zeta BOO (9)	14 41 +13 44	DS	4.5, 4.6 / A3IV, A2V	0.9"	Bright white, matched tight duo. Presently test for a good 5-inch - an elongated egg in 3-inch glass at high X. A close binary with 125-year period. The stars were 1.2" apart in 1959 & are now closing to a minimum separation of 0.03" in 2021. Watch them slowly merge into a single image! "Great optical test object." With a 6-inch or larger scope, mask-down to find smallest aperture that still splits it.

Object	Coords	Type	Mag / Spectral	Sep	Description
39 BOO (6)	14 50 +48 43	DS	6.2, 6.9 F6V, F5V	3"	Tight nearly-matched pair. "A neat DS." "Pretty object - white & lilac" or "certainly purplish." "Fine color contrast." Split in 2.4-inch, but best in 4-inch or bigger.
44 BOO (8)	15 04 +47 39	DS	5.3, 6.2 G0V, G2	2"	Another of Bootes' amazing flock of doubles! Binary with 220-year orbit - will reach maximum gap of 3" in 2020. " Miniature of Castor." "Yellow, & bluish, ruddy or purplish. Great difference as to colours." "Both yellow." "Both white." "Remarkable & highly interesting star." B is an eclipsing binary with 6.5-hour period & 0.5-mag. range in brightness. Resolved in 3-inch but needs 6-inch for good view. D=43LY
Struve 1835 BOO (7)	14 23 +08 27	DS	5.1, 7.4 A0V, F3V	6"	Neat Struve pair for small glass - easy in 2-inch. "Beautiful." "Pretty pair - flushed white, smalt blue." "White & lilac...sometimes blue, more usually tawny...an uncertainty of hue which I have found troublesome...in (companions of) some pairs."
Delta BOO (5)	15 16 +33 19	DS	3.5, 7.4 G8III, G0V	105"	* Two great neighboring examples of the many DSs in the sky too wide for typical big scope fields to be attractive but which are neat sights in small, low X/WF instruments.
Nu BOO (1)	15 31 +40 50	DS	5.0, 5.0 K5III, A5V	15'	Delta: "Bright yellow, fine blue...among pairs & triplets." "So easy an object." Nu: "Fine wide pair." "A lovely & equally bright double - orange & blue-white."
32 CAM=Struve 1694 (7)	12 49 +83 25	DS	5.3, 5.8 A1III, A0V	22"	! Lovely but sadly neglected, matched off-white combo, easy in the smallest of glasses. "Pale yellow & pale lilac or pale violet." "Neat DS - both bright white." Far-N (just 7 degrees from Polaris) CPM pair isolated from main body of constellation, resulting in undeserved obscurity. Sweet in 4-inch at 45X. Well worth the search! D=495LY
U CAM (0)	03 42 +62 39	SS	8.1-8.6 N7.7	---	A glowing stellar ember! "Reddest star in the sky." "Extraordinarily red." Rather faint, yet visible in binoculars - its appeal grows with telescope size. Best seen with good transparency & at least a 6-inch scope. Use direct vision to view its vivid color.
NGC 2403 CAM (5)	07 37 +65 36	SG	8.4	18'x11'	! "One of brightest galaxies & finest spirals in N sky - magnificent." "A prize object." "Presents an ever-changing vista as aperture increases. In 4-inch...a lovely gem. An ocean of turbulence & detail in 10-inch." Visible in binoculars as "A large hazy spot." One of biggest & brightest non-M galaxies. "It's too bad Messier missed this spiral" for it would be much more widely-known! "Easily seen elliptical glow" in a 4-inch. "Structure greatly resembles M33 (TRI)." Mottled texture visible in 8-inch & larger telescopes at medium X. "A lovely gem...an ocean of turbulence." D=12,000,000LY
NGC 1501 CAM (4)	04 07 +60 55	PN	11.9/14.0	55"x48"	"Small, curious, with star center." "An-out-of-the-way PN." "Somewhat like a flower." "A perfect little planetary." "Bluish-white tint - curious body." "Uniform light & definition abrupt." "Hazy smudge. Small bright smoke ring with faint central star." Dim bluish-grey disk - dark center/ring-like structure visible in larger scopes.
NGC 1502 CAM (2)	04 08 +62 20	OC	5.7	8'	Small clan of 60+ stars located just 1.5 degrees N of 1501. "Delicate 'Golden Harp' cluster." "Striking trapezoid-shaped cluster." "Beautiful & brilliant field...many

pairs." Includes the multiple stars Struve 485 (7, 7, 10, 10 at 18", 70", 139") & Struve 484 (all 10th-magnitude at 5", 23", 49"). "A small celestial fireworks display." "Easy to see...hard to locate." Best viewed in medium-size apertures. D=3,100LY

(Please see Page 95)

Object	Coords	Type	Mag / Spectral	Sep	Description
Zeta CNC (17)	08 12 +17 39	DS	5.6, 6.0, 6.2 F8V, F9V, G5V	0.9", 6"	! Beautiful close matched trio, all yellow. Two stars in 2-inch, three in 5-inch. "A splendid sight." "Remarkable phenomenon." "Famous triple system." Separation of A-B pair ranges from min. of 0.6" to max. of 1.2" (reached next around 2020). Forms an "excellent test" but is "tough." Orbital periods 60 & 1150 years. D=70LY
Phi-2 CNC (7)	08 27 +26 56	DS	6.3, 6.3 A6V, A3V	5"	Little-known, cozy matched-white combo. "Elegant." "Very pretty." "Close...both silvery white." Split even in 2-inch, neat sight in 3-inch. Stars related as a CPM duo.
Iota CNC (17)	08 47 +28 46	DS	4.2, 6.6 G8III, A3V	30"	! **Spring Albireo** Superb orange & blue pair resembling Albireo (Beta CYG)! Easy in a 2-inch glass & just resolved in binoculars. "Beautiful contrast." "Pale orange & clear blue." "Crocus & violet." "Wide...nice." "Striking." CPM system. D=165LY
57 CNC (8)	08 54 +30 35	DS	6.0, 6.5, 9.1 G7III, K0	1.4", 56"	Good close test double with much wider third star. "Both crocus yellow." "One of the loveliest pairs." "Like one elliptical star." "Exquisite object." Split in 4-inch.
X CNC (0)	08 55 +17 14	SS	5.6-7.5 N7.7 (C5)	---	! A stellar ruby! One of the reddest stars in *Sky Cat 2000.0.* "Very red." Tint obvious even in a 3-inch glass. As is the case for all color observations, stare directly at the star to see its hue (as opposed to averted - or side - vision, used for viewing faint sights like nebulae, galaxies, & dim stars in open & globular clusters). SE of Beehive (next).
M44 CNC (20)	08 40 +19 59	OC	3.1	90'	! **Beehive Cluster** A sprawling commune of more than 50 stars. "Just resolved by the naked eye; too large for usual (telescopic) fields, but full of fine combinations." "Hazy heart of Cancer." "Historic Praesepe." The "Cloudy One", "Little Mist" & "weather forecaster" of the ancients. "Very large & brilliant." "One of finest OCs in the sky." "Magnificent swarm of stars - huge." "Many doubles, triples & multiple stars of varied hues." "Two triangles will be noted." "Naked-eye gems." Tints of brighter members evident. "Exciting object" in binoculars & WF scopes but can be disappointing telescopically. "Ghostly sheen of cobwebs." "Easily visible to naked eye as a fuzzy glow...excellent views provided by 7x50 binoculars & RFTs. Too sprawling a group to give good views in 4-inch even at 25X (?) but does show well the colors of the brighter stars." "Becomes a dull object in large scopes." Galileo saw 36 stars here. "Symbolic of Spring. Brilliant show object." "A diamond mine." One of nearest: D=590LY
M67 CNC (17)	08 50 +11 49	OC	6.9	30'	! A beautiful but often-overlooked cluster in the shadow of the Beehive. "Vivid rich type of its exhaulted class." "Splendid." "Stunning." "Resembles a nebula in small instrument." "Beautiful object - marvelous contrast to Beehive." "Whorls of stars remind one of a whirlpool." "Magnificent open star cluster." "Partly encircled by brighter stars." "Star hues are predominantly rust, orange, gold & yellow." Some 200

stars visible with an 8-inch; over 500 actual members in this stellar clan. Nice sight in all apertures. One of the oldest OCs known - age 10 *billion* years! D=2,500LY

Alpha CVN (20)	12 56 +38 19	DS	2.9, 5.5 B9-A0p, F0V	20"	**! Cor Caroli** Striking blue-white pair in even the smallest glass. "One of the prettiest doubles in the sky. Golden yellow & lilac." "A blaze of glory." "Flushed white, pale lilac." "Pale yellow & fawn." "Creamy white, fine blue." Other tints reported for the fainter star are tawny, pale copper, pale orange, pale ruddy & pale olive blue! "At high X, the (companion's) purple (tint) is undeniable." "Subtle shades." "One of the most attractive DSs for the small telescope." CPM duo - "Definitely form a physical pair." One of the finest of its class! Primary is a strange "magnetic variable" star. D=130LY
2 CVN (10)	12 16 +40 40	DS	5.8, 8.1 M1III, F7V	11"	"Gold & blue; great color contrast!" "Striking though not conspicuous." Red & blue, orange & blue, gold-yellow & azure, golden-yellow & smalt blue, & yellow-white & rosy also given for its tints! Needs at least 6-inch to appreciate. Lovely in 10-inch.
Y CVN (8)	12 45 +45 26	SS	5.5-6.0 N7.7 (C5)	---	**! La Superba** Fiery carbon star. "Wonderfully red-orange & beautiful!" "Brilliant red color...noted for splendid color of its flashing rays." "Deep rich orange in 6-inch, a fiery coal. Color even redder in finder or binoculars." "Blood-red." "Deep red." "One of the most colorful stars in the sky." Looks orange with red tinge most nights. Spectrum displays an "Extraordinary vivacity of its prismatic rays...dazzling zones... broad spaces of profound obscurity." Diameter nearly one *billion* miles! D=400LY
M3 CVN (21)	13 42 +28 23	GC	6.4	16'	**! Spring's Globular** - 1st bright one! "Beautiful...superb...splendid...incredible swarm of countless stars, massing to a wonderful central blaze with glittering streams of stars running out on all sides." "Pre-eminent in its class." "A noble object...gorgeous mass of stars, blazing splendidly...running up into a confused brilliancy towards the centre." "A grand phosphorescent star-mass." "An amazing display." "Beauty rivals M13." "Stunning." "Spectacular." "I stood in awe of M3's vast number of stars." "A grand sight! Grainy texture. Tightly packed stellar swarm." Resolved across its brilliant nucleus in 8-inch at 160X - an awesome ball of stellar bees in 14-inch! D=35,000LY
M51/NGC 5195 CVN (20)	13 30 +47 12	SG PEC	8.4 9.6	11'x8' 5'x4'	**! Whirlpool/Question-Mark Galaxy** Big, beautiful face-on spiral with peculiar little interacting companion. "Magnificent object...finest of its class...central condensation, distinct spiral arms...amazing outer universe." "Transcendent 'Whirlpool Nebula' of Lord Rosse, resembling more an eternal question-mark - a supernal celestial enigma which in very truth it is." "Wonderful spiral." "Faint milky radiance around star-like nucleus." "Remarkably singular double nebula, the larger ring-shaped." It's "Seen exactly broadside." "Renowned. Classic example of a face-on spiral galaxy...pinwheel structure." Visible in 3-inch as "two very unequal neb. nearly in contact" & "impress-ive" in a 4-inch, M51 is typically disappointing in less than a 6-inch. The spiral arms can be seen in an 8-inch with averted vision on a dark night. Yet it took Lord Rosse's mighty 72-inch metal-mirrored reflector to first recognize them! This fact provides a

perfect illustration of Sir William Herschel's famous dictum that "When an object is once discovered by a superior power an inferior one will suffice to show it afterwards"! "This is the finest object of its class, but is always very disappointing to those who have seen only its photograph...because it is so faint. Nevertheless the spiral structure can without much difficulty be distinguished in the telescope." Actually, a marvelous sight in 12-inch & larger scopes! "Famous...one of the finest objects in the heavens." The Whirlpool can be glimpsed in binoculars under dark skies. D=36,000,000LY

M63 CVN (10)	13 16 +42 02	SG	8.6	12'x8'		! **Sunflower Galaxy** "Hallmark of Spring sky that should not be missed...a memorable sight in large backyard instruments." "Like an unresolved GC." "Very impressive in small scopes. Has strange visual appearance - one end more pointed than the other & oval grainy-looking central condensation." "Multiple-arm spiral...resembles some vast celestial flower." A bright milky-white glow with mottled structure visible in 4-inch at 100X or more on a good night. Appears as a "Silvery sliver of light" in 2- to 3-inch glasses. Striking in 8-inch - beautiful sight in 13-inch at 145X! D=35,000,000LY
M94 CVN (13)	12 51 +41 07	SG	8.2	11'x9'		! Small tightly-wound spiral, easy to find & see in even the littlest of scopes. "Very bright & comet-like." "Like an 8th-magnitude comet." "A grand object! Rapidly brightening toward a brilliant center which doesn't look like a star." "Very compact, nearly circular...remarkable for intense brilliance of its central core." "Much like an unresolved globular cluster." Bright but featureless in 3-inch glass. D=22,000,000LY
M106 CVN (9)	12 19 +47 18	SG	8.3	18'x8'		! Another easy spiral for small glasses. "Large white nebula...a noble-sized oval with brightish nucleus." "Large pear-shaped." "Whopping" apparent size & "remarkably high" surface brightness. "Big & bold in 6-inch - a grand galaxy to contemplate." "Grand object." It, like "All galaxies deserve to be stared at for a full 15 minutes" to appreciate their visual wonders & cosmic significance. D=33,000,000LY
NGC 4244 CVN (4)	12 18 +37 49	SG	10.2	16'x2'		A somewhat fainter version of famed NGC 4565 in COM, but without its dust lane. "Large distinct edge-on spiral." "Extremely long, narrow ray." Best in 8-inch & up.
NGC 4631 CVN (10)	12 42 +32 32	SG	9.3	15'x3'		! **Humpback Whale Galaxy** A big bright edge-on spiral. "Very long ray." "Long, arrowy; possibly two joined." One of the largest of its class, but missing usual central dust lane. "Mottled edge-on with close companion" - tiny 12th-mag. EG NGC 4627, lying off center. Name comes from its unusual 'lumpy' shape! D=39,000,000LY
Alpha CMA (14)	06 45 -16 43	DS	-1.46, 8.5 A1V, WDA	2.5"-11"		! **Sirius** The "Dog Star." Also, "The Nile Star", "The Sparkling One" & "The Scorcher." A blazing blue-white diamond-burst with famed white-dwarf companion! "Leader of the host of heaven: a glorious object." "Splendid...brilliant...truly dazzling." "Intense white with a sapphire tinge." "Crown jewel of CMA." Its tiny companion is now opening from its last minimum separation in 1993 to its next maximum in 2022. Orbital period=50 years. Has been seen with 6-inch - currently

needs optically-superb 12- to 14-inch, excellent seeing conditions & positioned on or near the meridian. "Frightfully difficult." "An object glass of 6-inches one night will show the companion to Sirius perfectly: on the next night, just as good in every respect, so far as one can tell with the unaided eye, the largest telescope in the world will show no more trace of the small star than if it had been blotted out of existence."! So repeated observation is key to seeing it. William Herschel stated that Sirius entered the eyepiece field of his huge metal-mirrored, home-made 48-inch reflector "with all the splendour of the rising Sun, & forced me to take the eye from that beautiful sight."! "Splendid...dazzling...brilliant white with definite tinge of blue." The ancients called Sirius "redder than Mars", "fiery red" & "classed it with Antares in hue."!? "Its colour has probably changed." A study done over a century ago "Established beyond doubt the ancient redness of the star." One fascinating possibility is that the now tiny white dwarf companion was in the red-giant stage back then & outshone Sirius itself! Current theories of stellar evolution say this can't happen in so short a time - but can we really be sure about this? "Was fiery-red: it gradually faded to a pure white, & is now a decided green."!? "Entrancing, particularly viewed as it is rising or setting" at which time it "Seems to change colors constantly & kaleidoscopically" - flickering "All the colors of the rainbow." Nearly twice the size of the Sun & 23 x its luminosity. Closest of all the naked-eye stars visible from mid-northern latitudes: D=8.6LY!

Epsilon CMA (6)	06 59 -28 58	DS	1.5, 7.8 B2II, B6	8"	
Herschel 3945= 145 CMA (5)	07 17 -23 19	DS	4.8, 6.8 K3I, F0	27"	
M41 CMA (19)	06 46 -20 45	OC	4.5	38'	
NGC 2360 CMA (6)	07 18 -15 37	OC	7.2	13'	

Adhara Ideal "warm-up" for Sirius! Should be easily split if you hope to glimpse its dazzling neighbor's companion. "Brilliant pair." "Pale orange & violet." Both look blue-white to most eyes. Needs at least 6-inch & still night for good view. D=490LY

! Winter Albireo A little-known, but magnificently-tinted easy combo! "Orange & blue, like Albireo." "Fiery red & greenish blue." "My favorite double." "Beautiful pair." "High yellow, contrasted blue." "Splendid colors." Its intense hues are striking in a 3-inch at 30X & are absolutely superb in a 6-inch at 50X! A "must see" wonder!

! Lovely big, bright coarse group of 80 gems 4 degrees below Sirius. "Sparkling." "Superb." "A splendid low-power sight." "A grand view & indeed one of the finest OCs for very small apertures. Brighter members form a butterfly, but the cluster as a whole is circular with little concentration." "Truly imposing...ruby central star." "Red star near center shows clearly." "Stars dividing into groups & curves...called superior to the Perseus Clusters." "Sapphires against black velvet." "Cluster bursts with splendor." "One of the sky's grandest OCs for binoculars. Fills a low-X field with shimmering stars." "Pre-eminent." A lovely sight in 4-inch at 45X. Visible to the unaided eye on a dark, Moonless night. "This mighty galactic cluster." D=2,400LY

"Beautiful cluster...melting into a very rich neighbourhood, as though the Galaxy were approaching us." "Singular group of very lucid specks." A large rich clan of 60-some sparkling gems in a "Winter Wonderland" MW field. Very pretty in a 5-inch at 50X.

Name	Coordinates	Type	Mag	Sep/Size	Description
NGC 2362/Tau CMA (6)	07 19 -24 57	OC	4.1	8'	**! Tau Canis Majoris Cluster** Seldom-observed glittering jewel-box of 60 diamonds surrounding the 4th-mag. blue giant Tau (a cluster member). "Striking!" "Beautiful." "Unusually attractive." This compact group needs at least 6-inch to appreciate its rare beauty. "A starburst of fireworks frozen in space." Very young cluster. D=5,400LY
				(Please see Page 95)	
Alpha CMI (2)	07 39 +05 13	SS	0.38 F5IV	---	**Procyon** The "Little Dog Star." "Closest of the high magnitude stars save Sirius." "A splendid star." "Brilliant." "A fine pale yellow star." Like Sirius, has white-dwarf companion in orbit around it - discovered visually in the Lick 36-inch refractor & very hard to see (mag. 10.8, ranging from 2" to 5" separation in 40-year period). "Difficult! Tougher than (seeing) Sirius & (our) Sun!" at the same time! Nearby: D=11.4LY
Alpha-2 CAP	20 18 -12 33	DS	3.6, 11.0 G9III	7"	**! Algiedi** Striking widely-spaced (378" - or 6.5' apart) orange pair for the unaided eye & binoculars - probably the finest of its kind in the sky! Also a magnitude-contrast "double-double" system for the telescope! "Magnificent!" "Noble pair, obvious to naked eye." "A fascinating hybrid system." The companion of Alpha-2 (the brighter of the two) is itself double - mag. 11.3 at 1.2" separation. Surprisingly, Alpha-2 & Alpha-1 are only an unrelated optical pair: D=100LY & 500LY, respectively!
Alpha-1 CAP (14)	20 18 -12 30	DS	4.2, 9.2 G3I	45"	
Beta CAP (10)	20 21 -14 47	DS	3.4, 6.2 G5II or K0II +A0V, A0III	205"	Another roomy, easy binocular-telescope duo. "Fine color contrast." "Very yellow, blue." "Orange & blue field-glass double." "Orange-yellow, sky-blue." Primary a complex spectroscopic triple system! "B also schizoid; 0.8", 6th- & 10th-mag., tough in 8-inch but not impossible." AB a CPM pair. Nice in 2-inch at 25X. D=560LY
Omicron CAP (9)	20 30 -18 35	DS	6.1, 6.6 A3V, A7V	22"	Well-spaced, closely-matched blue-white double for small glass. "Striking." "White & bluish - a pretty pair." "Both bluish." A really neat sight in a 2.4-inch at 30X.
M30 CAP (11)	21 40 -23 11	GC	7.5	11'	Little-observed GC due to its low DEC. "Moderately bright; beautifully contrasted with 8th-mag. star beside it; comet-like with 64X." "Surpassing cluster." "Pale white & fairly bright, elliptical with a 'central blaze'." "Splendid object...visual appearance quite unusual for a GC." "Beautiful." "Elliptical aspect." Needs aperture - nice sight in 10-inch on dark steady night. "Not one of the great globulars."! D=40,000LY
				(Please see Page 95)	
Eta CAS (19)	00 49 +57 49	DS	3.4, 7.5 G0V, K7V or M0V	13"	**! Easter Egg Double** Beautiful yellow & ruddy-purple pair. A "Topaz & garnet"! Companion also called lilac, lavender, red & orange. "Superb." "Like a balloon & its gondola." "Especially beautiful contrast in colors." Some early observers reported tints as "orange & gold" & "red & green"?? The magnitude contrast adds to appeal of this lovely combo. Neat in 3-inch glass - stunning in 6-inch. Orbital period 480 years; the stars were 5" apart in 1889 & are now slowly opening to 15" in 2125. D=19LY
Iota CAS (17)	02 29 +67 24	DS	4.6, 6.9, 8.4 A5, F5, G5	2.5", 7"	**!** Tight triple system. "Impressive little trio!" "Golden yellow, lilac & purple." "A splendid triple star." "Yellow, lilac, blue - fine but not easy object." "Challenging." "Yellow, blue, blue...very fine." "Beautiful triple." "Elegant." "One of the finest

triple stars that the sky has to offer." Tight for 4-inch, easy in 8-inch with good seeing. Orbital period of close duo 840 years, opening to 2.7" in 2103. All CPM. D=160LY

Sigma CAS (12)	23 59 +55 45	DS	5.0, 7.1 B1V, B3V	3"		

(Please see Page 95)

! Close blue-white & green-white pair for medium-size telescopes. "Green, very blue." "Color intense...grand field." "Glorious low-powered field." "Teeming with jewels." Tints are obvious in 5-inch at 100X & a vivid sight in 10-inch at 160X! "In a superb region." "Beautiful - flushed white, smalt blue; the colours are clear & distinct." Split in 3-inch, impact grows with aperture! "Not very remarkable."? Fixed. D=1,400LY

Struve 163 CAS (6)	01 51 +64 51	DS	6.8, 8.8 K5	35"

Dim but colorful double. "Red-gold, blue. Colours splendid." "Remarkable colors." "Ruddy orange, blue." Also 9.7-mag. companion at 115". Color contrast best-seen in 6-inch & larger instruments. Just one of the many faint deep-hued Struve pairs in sky!

WZ CAS (1)	00 01 +60 21	DS	7.6-10, 8.7 N1, A	58"

Wide, faint but striking red & blue pair. Primary a red carbon variable with 186-day period. Beautiful sight in 6-inch at low X when near its maximum light, yet virtually unknown as a pretty DS! "Fine colors...wide double." "Remarkable" in an 8-inch.

M52 CAS (14)	23 24 +61 35	OC	6.9	13"

! Rich sparkling clan of at least 100 suns. "Beautiful sight in small scope. Distinctive pattern: a needle-shaped inner region inside a half circle." "A most beautiful cluster, somewhat triangular." "An 8th-mag. topaz star" embedded within its apex, "as is frequently the case." "Extended center, fairly clear." A lovely "Bird-in-flight group... somewhat like popular M11 (SCT) but less bright." "A bird with outspread wings." "Of singular beauty." "One of the richer & more compressed OCs." Pretty in all size scopes. D=4,000LY The very faint **Bubble Nebula** (NGC 7635) lies just 36' to SW.

M103 CAS (8)	01 33 +60 42	OC	7.4	6'

Arrowhead or fan-shaped small group of several dozen stars. "Vivid cluster in grand setting." "Beautiful field...containing Struve 131 (7, 10, 11 at 14" & 28") & red star." "Central diamond shape." "A beautiful 10th-mag. reddish star prominent, its colour rose-tinted." "Not one of the richer clusters but...fairly compact...(with) wedge-shaped outline." "Dominating the cluster but not a member is the DS Struve 131. A grand view! The stars form an arrowhead...many of the fainter stars colored." The last entry in Messier's original classic list. Attractive in a 4-inch at 45X. D=8,000LY

NGC 281 CAS (2)	00 53 +56 36	DN	7.8	35'x30'

* A typical big, subtle & overlooked DN. "Large faint nebulosity near Eta CAS." A "Complex glow with a handful of stars." "A large, faint, amorphous mist surrounding the multiple star Burnham 1...a triangular blur with rounded tip & a brighter region." The imbedded DS consists of an 8th-mag. primary & three fainter companions of 9th- to 10th-mag. The nebulosity & stars are visible in a 5-inch, but at least a 10-inch is needed to really appreciate this interstellar fog. Also cataloged as an OC in the NGC?

Phi CAS/NGC 457 (10)	01 19 +58 20	OC	6.4	13'

! **Owl/ET Cluster** A striking group of some 80 suns arranged in the shape of an owl, with golden 5th-mag. Phi & a 7th-mag. star (both supergiants) marking its bright eyes.

"A very elegant group attending Phi." "One of best CAS clusters." "Massed jewels! Topaz Phi close following." "Splendid...rich scattered group of stellar points...true splendor." "Stars spread out like wings...one of the best...that Messier missed." A "Beautiful sight...particularly pretty." "Delightful OC...so easy to find!" A 3-inch at 45X shows it nicely - spectacular in an 8-inch at 60X. "A grand sight." D=9,300LY

NGC 663 CAS (9)	01 46 +61 15	OC	7.1	16'	

"Rich cluster with several DSs." These combos are Struve 151, 152 & 153 - all 9th- & 10th-mag. pairs, each roughly 8" apart. "Neat DS (153) in a cluster...in an elegant field...down to infinitesimal points." "Reveals only a few points of light against an unresolved mass" in small scopes. "Fine triple cluster." Refers to the small, dim clumps of stars NGC 654 & NGC 659, both lying within a degree of NGC 663 itself.

NGC 7789 CAS (16)	23 57 +56 44	OC	6.7	16'	

! A big, even sprinkling of more than 300 faint stars! "Remarkably rich & uniform." "Exceptionally rich galactic star cluster...at least 1,000...actual members." "Radiant." "A most superb cluster of stars & stardust!" "Beautiful large faint cloud of minute stars." "A mere condensed patch in a region of inexpressible splendour...a cluster of minute stars, on a ground of star-dust...a very glorious assemblage, both in extent & richness, having spangly rays of stars which give it a remote resemblance to a crab... beautiful object." "Magnificent...often neglected." "Large & condensed!" A "Gentle collective blur floating amid a field strewn with stardust." "Nearly 1,000 stars - most have evolved into red giants & supergiants - cluster is well over a billion years old." "One of the densest OCs...almost 600 stars all crammed into just 16'." "View grows steadily more impressive with every increase in telescope size." Visible in a 2.4-inch & magnificent in 6-inch or larger scope. "Splendid swarm." Angular size seems more like 30' across. Intermediate in concentration between rich OCs & loose GCs - perhaps a 'semi-globular'? Sometimes called "Caroline's (Herschel) Cluster." D=6,000LY

Omega CEN (15)	13 27 -47 29	GC	3.6	36'	

! **Omega Centauri Cluster** Although 2.5 degrees under survey's -45 deg. limit, "This most glorious object" simply *must* be included in any listing of top celestial wonders! Largest, brightest (so bright that it was assigned a Greek letter by Bayer as a naked-eye star!) & finest of all the GCs, it's a wondrous sight as seen from the southern states - & especially from the Florida keys. One well-known deep-sky observer of the past once claimed that Omega was more beautiful through a pair of binoculars than was famed M13 (the Hercules Cluster) in his 10-inch reflector! "Magnificent swarm of countless suns - resembling a vast swarm of bees." "Noble GC - beyond all comparison the richest & finest object of its kind in the heavens...the stars are literally innumerable." "Grandest of all GCs." "Humongous!" "Few celestial sights can compare with its raw beauty. Your first sighting...will be remembered for a lifetime." "One of the most magnificent objects within range of the telescope." "Enthralling." Over degree across on photos - contains at least a million suns! Atmospheric absorption, haze & clouds near the horizon all greatly affect the visibility (& impact) of both this object & the one that follows, as viewed from mid-northern latitudes. Truly awesome! D=17,000LY

NGC 5128 CEN (10)	13 26 -43 01	SG+EG? 7.0	18'x14'	**! Black Belt/Karate Galaxy** "At first glance...looks like two separate objects; closer scrutiny shows large glow with equatorial dust lane." Peculiar system - possibly two colliding galaxies - an elliptical & spiral merged! "Strange system...subject of much controversy." Very low from mid-northern latitudes - needs at least a 10-inch scope & dark sky to see black central belt splitting the galaxy. Visible in binoculars, but only "an amorphous sphere" in the telescope on a typical night. An intense source of radio emission, it carries the designation "Centaurus A" - which is the name by which most professional astronomers refer to this enigmatic object. D=22,500,000LY
Beta CEP (15)	21 29 +70 34	DS	3.2, 7.9 B2III, A2V	13"	**!** Neat unequal pair, both blue-white - exquisite even in a 3-inch. "Greenish white & blue or purple." "Beautiful object." Primary short-period Cepheid variable. D=980LY
Delta CEP (17)	22 29 +58 25	DS	3.5-4.4, 6.3 F5I-G1I, B7	41"	**!** Striking, wide pale-orange & blue duo. "Especially fine pair, somewhat like Beta Cygni (Albireo)." The "colours being in fine contrast." "One of the most beautiful DSs in the sky - blazing yellow, bluish-white." "Beautiful fieldglass double." "Two lambent clusters in field." "Celebrated." Primary is prototype of the famed pulsating Cepheid variables used as intergalactic distance indicators - period 5.4-days. Sadly, it's often overlooked as an attractive DS due to its notoriety as a variable! D=1,300LY
Krueger 60 CEP (3)	22 28 +57 42	DS	9.8, 11.3, 10.1 M3V, M3V	2.5", 75"	***** Famous red-dwarf binary & flare star with 44-year orbital period. In same wide field with Delta CEP, 43' to its SW. "Faint but remarkable double." "Both red!" "The chance of detecting a flare (B component: several magnitudes, lasting several minutes) adds to the interest in observing this unusual system." A good 8-inch at high X can split the close pair on a steady night. In smaller scopes, it's an elongated egg whose axis can be seen turning over the course of just a few years from the very rapid orbital motion of the stars! The C component is only optical & unrelated. A "Bevy of 10-through 14-mag. stars within 1' radius all associated with this...system." One of the nearest of all the visual binaries to us: D=13LY
Xi CEP (11)	22 04 +64 38	DS	4.4, 6.5 A3, F7	8"	**!** Bright easy pair for small glass - delicate "subtle" color contrast. "White, tawny or ruddy." "Splendid DS...both bluish." "Blue-white & yellow." Nice in a 2.4-inch at 30X - yet tight enough to still remain attractive in a 13-inch at 145X. D=80LY
Omicron CEP (6)	23 19 +68 07	DS	4.9, 7.1 K0III, F6V	3"	Tight pair for larger scopes. "Elegant...orange yellow, deep blue; the colours in fine contrast." "Gold yellow & azure." "Very yellow, very blue." "Primrose (!) & lilac." "Yellow, yellowish-green; not remarkable as a contrast." Needs high X, good seeing.
Mu CEP (9)	21 44 +58 47	SS	3.4-5.1 M2I	---	**! Herschel's Garnet Star** "Reddest naked-eye star in N sky." "Shining like a drop of blood." "Deep orange." "A striking deep-orange or red beacon amid an infinite sea of stars." "Curious." "Beautiful deep red." "Very red." "Striking tint...notable in binoculars." "Usually appears a deep orange-red, but on occasion seems to take on a peculiar purple tint." "A most beautiful object." Mu's color appears to depend upon

aperture - looks almost red in a 3-inch, deep orange in an 8-inch and yellowish-orange in a 13-inch. A huge supergiant - one of the very largest stars known in the universe (possibly as much a *billion* miles across!). Yet cool enough that water vapor (steam!) exists in its outer atmosphere. Dubbed "Garnet Sidus" by Sir William. D=2,800LY

Struve 2816/IC 1396 CEP	21 39 +57 29	DS/	5.6, 7.7, 7.8 O6	12", 20"	! "Spectacular triple." "Fantastic, beautiful triple" system set within huge dim nebula & big faint starry grouping. "Emission nebula & scattered open cluster. Large wreath-shaped glow almost totally ignored by amateurs. Magnificent object."?? "Like a very faint Rosette Nebula." "Extremely large & diffuse area." "Large haze." The faint pair Struve 2819 (7.5, 8.5, 12") also lies in field! Overall complex best seen on dark night in RFT. (Showpiece is the triple itself.) Just 1.5 degrees S of Mu CEP. D=2,600LY
(6/1)		DN/ OC	--- 3.5	170'x140' 50'	
Struve 2840 CEP (6)	21 52 +55 48	DS	5.5, 7.3 B6, A1	18"	Another of Cepheus' many lovely doubles! "Greenish-white, bluish-white...splendid pair." "Very fine...B possibly purple." "Beautiful." Nice split in 2-inch glass at 25X.
NGC 40 CEP (4)	00 13 +72 32	PN	10.2/11.6	60"x40"	"Unusual red (?) planetary." "Bigger version of the Blinking Planetary (NGC 6826 in CYG)." "Central star in easy halo." "One of finest PNs at high declination - central star easily spied." "Greyish disk." In a relatively sparse region & not easy to find - but well worth the search! Visible in 5-inch at 50X positioned between two faint stars. Needs at least 8-inch & 100X or more to appreciate. "Rather difficult." D=3,000LY
NGC 188 CEP (3)	00 44 +85 20	OC	8.1	14'	* Oldest known OC & one of oldest objects in the universe - age 12 to 14 *billion* years! "This ancient one." "Immensely ancient." A "Ghostly glow." "Large but dimly luminous spot with only a few of the brighter members showing individually." Not easy even in a 10-inch - seems *much* fainter than published magnitude. Contains over 100 very dim stars. Just 4 degrees from Polaris, the Pole Star. D=5,000LY
NGC 6939/NGC 6946 CEP	20 31 +60 38	OC SG	7.8 8.9	8' 11'x10'	Unique cluster-galaxy combo, in same eyepiece field 38' apart! Close together in the sky but very far apart in space: D=4,000LY & 10,000,000LY, respectively! "Beautiful cluster, rich in faint stars." "Closely knit family of 80 stars crammed into a tiny 8' circle...one of richest clusters in N Autumn sky." "A grand but distant collocation of suns." "Magnificent face-on spiral" in photos, but quite faint visually due to its low surface brightness. "A mini M33." "No two ways about it...a tough catch." Cluster a relatively easy mist in 5-inch at 50X - galaxy a slightly more difficult glowing patch.
(3/2)					
				(Please see Page 95)	
Gamma CET (15)	02 43 +03 14	DS	3.5, 6.2 A3V, F3	3"	! Attractive tight duo with delicate tints. "Yellow, ashen, beautiful." "Curious colors." "Colorful pair." "Pale yellow, lucid blue, the colours finely contrasted - beautiful." Companion also described as lilac, olive green, ruddy, tawny & dusky! "These colors appear to be at least partly illusionary" based on the spectral types. Test for a good 2.4- to 3-inch glass at high X on steady night - not easy in 4-inch. "The two stars undoubtedly form a long-period binary" but the orbital period must be at least several thousand years! A 10th-mag. red-dwarf CPM companion lies 14' NW. D=63LY

Omicron CET (8)	02 19 -02 59	SS	2.0-10.1 M5III-M9III, B8	---	*** Mira/Wonder Star** Prototype & best-known of the long-period variable stars. "The quintessential example" & first of the variables to be discovered. "Majestic." "The Marvelous." "The Wonderful." "Symmetry in motion." "Amazing brightness range from mag. 2 to an out-of-sight mag. 10." "Very full ruby." "I found no trace of red."! "Always reddish." "Less red than gas flame." "Flushed yellow." "Classic. Due to temperature change, the red tint...slowly deepens as the star fades." "Extraordinary object." Pulsating giant & one of the largest stars known - 500 x the size of our Sun at maximum (& 250 x brighter)! Yet so cool that water vapor (steam!) exists in its outer atmosphere. Naked-eye sight at maximum light, telescopic at minimum over a 332-day period. Mira has a very close (1" distant) & faint, variable (9.5- to 12th-mag.) blue companion in orbit about it, circling the huge star every 400 years. D=220LY
M77 CET (11)	02 43 -00 01	SG	8.8	7'x6'	! A Seyfert galaxy - intense star-like core surrounded by circular haze. "Small, bright, wonderfully distant & insulated." "Easy target." "Faint but interesting." "Small, softly glowing...face-on spiral." An "Out-of-the-way galaxy...often bypassed...easy to spot." "Beautiful." Here's "One of the best galaxies for viewing in small apertures. Its irregular shape is beautiful at low to medium magnifications." Strangely, Messier & the Herschels thought this object to be a star cluster!? "Luminous knots & condensations...a mottled effect" in the spiral arms can be glimpsed in 6-inch & larger scopes. "An old friend...a marvelous sight (in 4-inch)...having a miniature quasar at its core." Other fainter galaxies lie nearby. Most remote of all the M objects: D=82,000,000LY
24 COM (17)	12 35 +18 23	DS	5.2, 6.7 K2III, A9V	20"	! Vivid orange & blue-green double with intense hues for small glasses. "Beautifully coloured." "Beautiful contrast." "Colours very brilliant...exquisite wide pair." Most see the primary as orange or rich yellow, while the companion has been described as aqua, emerald, blue, lilac & greenish-white. "Reminiscent of Albireo...slightly fainter but gorgeous." "Great color contrast!" In a 3-inch at 30X, "orange & emerald" seem to best describe this lovely tinted jewel! Colors still intense in 10-inch at 80X. Hidden amid the many faint lights of Coma Berenices - worth the hunt! CPM pair. D=300LY
M53 COM (10)	13 13 +18 10	GC	7.7	13'	"Brilliant mass of minute stars...blazing in centre...not very bright 3.7-inch; beautiful 9-inch." "A mass of minute stars & 'star dust'." "Interesting ball of innumerable worlds...wonderful assemblage...contemplation of so beautiful an object cannot but set imagination to work, though the mind may be soon lost in astonishment." "A superb object." "Very compressed." "Resolving into a wonderful swarm of tiny star images with larger instruments, partially resolved in 6-inch." "One of the most beautiful sights I remember to have seen in the heavens...a blaze of light." "Seems out of place in the Spring sky, stranded in the midst of the intergalactic furor of Coma Berenices." "Magnificent" in 12-inch. D=65,000LY 10th-mag. GC NGC 5053 lies a degree SE.
M64 COM (17)	12 57 +21 41	SG	8.5	9'x5'	! **Blackeye Galaxy** Superb bright spiral! "One of the grandest in the heavens - like a colossal pendent abalone pearl in rayless void." A "Magnificent large bright nebula

blazing to a nucleus." "A conspicuous nebula, magnificent in both size & brightness." "An oval glow with noticeably off-center stellar nucleus." "Dark splotch NE of core - beautiful at 200X" in 8-inch. "A vacuity below the nucleus." "Huge dark patch floats in front of half the galaxy's disk...(an) arc-shaped cloud of dust prominently visible in small telescopes." "Eye needs large aperture." It has been seen in a 2.4-inch; called "easy" in 4-inch glass but "subdued" in 8-inch. Actually anything over 6-inches shows it quite well. Strangely, this dark marking is not mentioned in the NGC's description of M64! "Because of its details, ranks as one of the finest Messier objects." And also as one of the easiest of all galaxies to see in small telescopes! D=25,000,000LY

M88 COM (11)	12 32 +14 25	SG	9.5	7'x4'	

"Bright multiple-arm spiral" with "luminous core." "In wonderful nebulous region. Marvelous as swept with 64X; identification difficult." "A vast Sargasso sea of star-illumined cosmic matter." "The voyager in this fabulous region may explore island universes by the dozens." "Pearl white in colour - spindle figure with its attendant stars forms a pretty pageant." "One of the notable spirals of the Coma-Virgo Galaxy Cluster." "Resembles a smaller Andromeda Galaxy. Grand in 4-inch. Core appears stellar surrounded by a soft glow." Easy catch in a 3-inch glass. D=40,000,000LY

M99 COM (8)	12 19 +14 25	SG	9.8	5'x5'	

Pinwheel Nebula Nearly face-on system for medium apertures. "Wonderful spiral." "Bright, circular, nucleus uncondensed...holds attention." "Large round nebula which, though pale, is well defined." "With vividly sparkling light." A "Triply branched spiral nebula...large, bright &...very remarkable." "An impressive sight." Typical of the amazing "profusion of bright nebulae" (galaxies) of every size, shape & type to be found strewn across this part of the heavens! Fascinating in 8-inch. D=50,000,000LY

NGC 4565 COM (11)	12 36 +25 59	SG	9.6	16'x3'	

! "Superb edge-on spiral with dust lane." "Beautiful...prominent dust lane." "Bright narrow streak in 6-inch - in 10-inch a perfect little needle of light." "Seen edgewise, with bulbous center and long 20' bifurcated luminous body - unique!" "Extraordinary phenomenon." "One of the sky's extragalactic treasures." "Classically elegant. Cigar-shaped body split by dark band & noticeable central bulge." "Long, peculiar object... appearing cut in two by a great dark band of obscuring matter along its edge...needs a 9-inch telescope to be seen to advantage." Agreed! Marvelous sight in a 13-inch at 145X! The dust lane can be glimpsed in a 4-inch on a dark night. D=20,000,000LY

MEL 111 COM (5)	12 25 +26 00	OC	1.8	275'	

* Coma Star Cluster The largest (& one of closest: D=270LY) naked-eye star cluster with some 80 members scattered over 5 degrees of the sky! "Now here is an OC that was made for binoculars!" But it's "Completely lost in the much narrower field of the telescope." To the eye a "Curious twinkling, as if gossamers spangled with dewdrops were entangled there...delicate beauty." "Coma Star Cloud." "6 degree diameter! Use binoculars for best view, but it's not too shabby a naked-eye treat as well." The wide DS 17 COM (5th- & 7th-mag., 145" apart) lies near the cluster's center. "A naked-eye treasure." "A shimmering haze...2X or 3X opera glass gives a wonderful view."

Gamma CRA (8)	19 06 -37 04	DS	4.8, 5.1 F8V, F8V	1.3"	Matched close pair of yellowish suns - binary with 120-year period. Stars were closest at 1.2" in 1992 & are now slowly opening to another maximum of 2.5" around 2060. "Fine binary." "Identical 5th-mag. yellow stars forming a tight double for small telescopes." Needs excellent seeing & at least 5-inch due to low altitude. D=69LY
Zeta CRB (15)	15 39 +36 38	DS	5.1, 6.0 B7V, B7V	6"	! Pleasing easy double with bluish-white & greenish-white tints. A "Fine", "Pretty", "Vivid " "Pair of blue stars." "White & turquoise; beautiful." "Bluish white, small blue." Fixed pair - hardly any change in position angle or separation of components in over a century, but they almost certainly do form a physical system. Very lovely in anything from a 2- to 14-inch telescope! Colors delicate but definite in all apertures.
Sigma CRB (10)	16 15 +33 52	DS	5.6, 6.6 G0V, G1V	7"	Another nice easy pair. "Wide, bright binary" with 1,000-year period, slowly opening from a minimum of 1" in 1828 to a maximum of 10" in the 24th century! "Great divergence as to colours." "Creamy white, smalt blue." "White & orange." "A pair of yellow stars for small telescopes." Companion appears "Sometimes blue, sometimes ashy, sometimes ruddy." "Certainly not blue."! What unearthly tints do you see here?
Eta CRB (6)	15 23 +30 17	DS	5.6, 5.9 G2V, G2	0.4"-1.1"	Very tight, rapid-moving binary system with 42-year period. Offers excellent test of telescope, atmosphere & observer. "Splendid DS but the pair are too close to be split by an aperture under 5-inches." Actually, they have been resolved in an excellent 3.5-inch Maksutov at less than 1" separation. "May be divided with high X on a good 6-inch throughout much of its orbit." Last widest in 1993. This "Wondrous physical object" has completed over four orbits since its discovery in 1826! D=50LY
R CRB (5)	15 49 +28 09	SS	5.7-14.8 F8p	---	* **Fade-Out Star** Amazing "reverse nova" - watch for its unpredictable disappearing act! "A remarkable star." "Puzzling behavior - bizarre instability." "One of the most remarkable variables in the heavens." The most "Famous of its ever-surprising type." "Merits careful attention." "Dimming caused by carbon soot condensing in a stellar wind." Fun to monitor: normally visible to the unaided eye & in binoculars - but can suddenly disappear, dropping more than 7 magnitudes & requiring a 12-inch scope to see! Months later, "R will brighten from the depths of obscurity" to its usual luster!
T CRB (4)	16 00 +25 55	SS	2.0-10.8 M3III+Bp	---	* **Blaze Star** Dramatic example of a "recurrent nova" & the opposite of R CRB - this strange star normally *stays* at minimum light & unpredictably flares up to maximum! "Another spectacular star - performs in almost the opposite way to R, usually slumbering around magnitude 11." The last recorded outburst was back in 1946, but "The observed maxima are so short (staying at peak-brightness for only a night or less) that it is quite possible that some have been missed...deserves continuous attention."
Delta CRV (14)	12 30 -16 31	DS	3.0, 8.4 B9V, K2V	24"	! **Algorab** Nice color- & magnitude-contrast duo. "Wide, lopsided pair, yellowish & pale lilac." "Blue-white & orange." "Pale yellow & purple." "Creme & lilac tints." "Yellow & pale violet." Truly "A wonderful DS when viewed through just about any

telescope." Some lists give the companion as faint as mag. 9.2, but it definitely seems brighter than this at the eyepiece. Pretty in a 3-inch at 45X, the companion's tint is very intense as seen in both a 10-inch at 80X & a 13-inch at 145X. CPM. D=125LY

Struve 1669 CRV (6)	12 41 -13 01	DS	6.0, 6.1 F5V, F3V	5"	Lovely matched yellow-white twins, easily resolved in a 2.4-inch glass at 60X. "Neat white-white pair." Often encountered in sweeping from Delta CRV to the Sombrero Galaxy (M104), across the border in neighboring VIR. Sweet duo in 4-inch at 45X!
NGC 4361 CRV (10)	12 24 -18 48	PN	10.3/13.2	80"	"Faint but even nebulosity around a star." "Round, with no sign of color...central star is visible with averted vision" in 8-inch. "Moderately large disk." "Small & bright." "One of the larger & brighter planetaries in the sky; has been spotted with binoculars." "Fluorescing ball of gas, bright blue disk." "Large but faint planetary inside the sail of Corvus." "Nebula's low contrast makes it extremely difficult in light-polluted skies." Some sources give its size as just 45" but that's only the inner core. "Small but really fine PN...circular object filled with mottled light" seen in 4-inch at 40X!? D=2,600LY
NGC 4038-9 CRV (2)	12 02 -18 52	SG?	10.7	3'x2'	**Antennae/Ring-Tail Galaxy** "One of the easiest colliding/interacting galaxies to observe - single nebulous arc in small scopes." "Shrimp-shaped." "Peculiar pair." "Two elliptical nebulae...in contact." "Two extending filaments." "One of the most unusual sights in the night sky." Quite fascinating in a 14-inch! D=90,000,000LY
Alpha CYG (2)	20 41 +45 17	SS	1.25 A2I	---	**Deneb** "Star of the Cross." 'Summer Triangle' asterism member along with Vega & Altair. "Brilliant white." "A super-sun!" Nearly as bright as Altair - but 100 times more remote (D=1,600LY)! "One of the greatest supergiants stars known...luminosity is computed to be about 60,000 x that of the Sun...diameter may be about 60 x...(our Sun) would appear 13th-mag. at its distance."! Blue-white hue seen in all apertures.
Beta CYG (21)	19 31 +27 58	DS	3.1, 5.1 K3II+B9V, B8V	34"	**! Albireo** One of the grandest sights in the entire heavens! Superb orange & blue pair in exquisite MW setting - finest colored double for small glasses. "Dazzling gold & deep blue duo." "Glorious contrast of yellow & blue-green stars." "Splendid fixed pair, superb color contrast...golden yellow or topaz & sapphire." "Golden & azure, giving perhaps the most lovely effect of color in the heavens." "An extraordinary pair of stars dazzling in their colors." "Thrilling pair!" "Brilliant...what an amazing DS." "Radiant stellar jewels." "Actually looks better in smaller scopes; a prime treat at the Swan's head!" "I have seen the colours beautifully by putting the stars out of focus ...they are actually different, not...from mere contrast" as can be seen by placing one of them out of the field or behind an occulting bar in the eyepiece - the other star "retains its own distinctive hue."! "The standard against which all other doubles are judged ...if you've ever doubted that stars have colors, this pair should remove any question." "One of the showpieces of the sky." "One of the most gorgeous... most observed DSs." "You're not a stargazer if you haven't seen this popular double." "Absolutely superb! Needs only 18X on 2-inch; likely a binary." "No more than 30X is required...to show

this superb pair as two contrasty jewels suspended against a backdrop of glittering star-dust." Actually a lot less power will do it - like many of the other doubles in the 30" & over separation range in this roster, it can be split with steadily-held 10X50mm binoculars! Albireo is in reality an incredibly tight visual triple system: A's composite image (note dual spectral type) has been resolved in a 26-inch refractor (1.5-mag. diff., sep. 0.4"). This close AB pair is some 400 *billion* miles from the C star - 55 of our solar systems placed edge-to-edge would fit in the 34" gap between them! D=400LY

Omicron-1=31 CYG (8)	20 14 +46 44	DS	3.8, 7.7, 4.8 K2II+B4V, B9, A5III	107", 338"

! Magnificent wide triple-sun system in radiant MW field - tints orange, blue & white! "Perhaps the most beautiful binocular double in the heavens, orange & turquoise, like a wider version of Albireo. Small scopes...show a closer blue companion." A "Fine wide color-contrast group." "Striking contrast, entire field exceptionally impressive, an extremely beautiful sight!" Both companions have been described as "cerulean blue", & the wider one as "white or yellowish, with an eye of blue (or) pale yellow with a cast of blue - a strange but accurate description." Trio superb in 3-inch glass at 30X! Primary is an eclipsing binary with 10-year period! D=200LY Omicron-1 has often been confused in lists & star atlases with nearby Omicron-2 (32) CYG. A vivid orange K-type giant sun, it's also an eclipsing binary - this one having a period of 3 years.

Delta CYG (12)	19 45 +45 08	DS	2.9, 6.3 B9III, F1V	2.2"

! Bright but very close, unequal pair for 6-inch & larger apertures - greenish-white & ashen. "Difficult - fairly severe test for 3- or 4-inch...companion lies virtually on the first diffraction ring of the primary & is thus often obscured except at times of very steady seeing." "Often easy in twilight, but invisible in a dark sky" due to glare from primary. "Notoriously difficult...for small scopes...beautiful." "Blue-green & blue." "A most delicate DS, beautiful...pale yellow, sea-green." Very slow binary: period 800 years, opening to 3" in 2225. Splendid sight in 13-inch refractor at 190X! D=270LY

Mu CYG (10)	21 44 +28 45	DS	4.8, 6.1, 6.9 F6V, G2V, A5	1.2", 200"

Tight 500-year-period binary closing to a minimum of 0.8" in 2017, teamed with wide optical member. A "Beautiful DS with distant companion...white, both blue...clear components." "Red-gold & purple." "Yellow, tawny or blue, lilac...colours variable?" Dynamic duo for apertures of 6-inches or more, plus a neat binocular combo. D=70LY

16 CYG (9)	19 42 +50 32	DS	6.0, 6.1 G2V, G5V	39"

Perfectly-matched, roomy golden pair - a lovely sight in small scopes! Also guide for locating the well-known Blinking Planetary (described below), which lies just 45' due E of the double. "Nice & easy pair, CPM." "Both pale fawn-colour." "Superb field."

61 CYG (17)	21 07 +38 45	DS	5.2, 6.0 K5V, K7V	30"

! **Piazzi's Flying/Bessel's Star** Beautiful, easy orange double! Famous long-period (650-year) binary - "The first of the host of heaven to reveal to Bessel (1838) the secret of their distance. How vast must be the dimensions of this great Universe! What a temple for the Creator's glory!" "That splendid result." One of the very closest stars known: D=11LY - or some 64,000,000,000,000 miles! This double's high annual proper motion of 5" means it moves the apparent gap between the two stars in just 6 years! Also a suspected planetary system! "Showpiece pair of orange dwarf stars." "A double orange." "Distinctive orange color...against the background blue & white stars of the MW...quite pretty." Striking in 3-inch at 30X; tints hold up in big scopes.

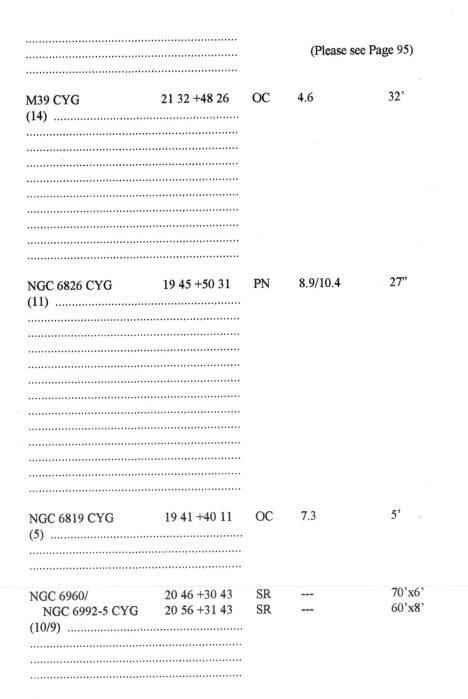

(Please see Page 95)

M39 CYG (14)	21 32 +48 26	OC	4.6	32'
NGC 6826 CYG (11)	19 45 +50 31	PN	8.9/10.4	27"
NGC 6819 CYG (5)	19 41 +40 11	OC	7.3	5'
NGC 6960/	20 46 +30 43	SR	---	70'x6'
NGC 6992-5 CYG (10/9)	20 56 +31 43	SR	---	60'x8'

! "Grand open type." "Too large & sparse a group" for usual telescopic fields - lovely in binoculars & RFTs! "Triangular with DS in center." "Forms equilateral triangle with bright star in each corner...a pretty sight in very small scopes - less impressive in...4-inch." "One of the finest OCs for binocular users" where it "Frequently exhibits a striking 3-D effect."! "Rather splashy Galaxy field...in a very rich vicinity." There's about 30 suns in this bright, loose-knit group. *Given very low X & WF, a telescopic showpiece!* One of the nearest OCs: D=890LY M29, Cygnus' other Messier OC, lies less than 2 degrees SE of Gamma CYG. It appears as a "Small, bright, fully resolved, trapezoidal" shaped knot or "stubby dipper" of a dozen or so stars "looking like a tiny Pleiades." Although not a showpiece, it's on (7) lists & still worth looking up if you're in the area. Perhaps a transition object between the multiple stars & loose OCs??

! **Blinking Planetary** An amazing object - one of the few celestial wonders that 'does something' while you're watching it! Named coined by the author in the August '63 issue of *Sky &Telescope*. Shows pale blue disk with obvious central star. "Somewhat resembles a star out of focus." Staring right at its nuclear sun, the nebula itself turns completely invisible. Switching to averted vision, it suddenly reappears & swamps the star with its bluish radiance! Alternating between direct & averted vision creates a spectacular blinking effect. "Winks out due to foveal insensitivity." "Here again, gone again effect." "The result of a strange combination of eye, telescope & photons that will leave you remembering (it)." Effect visible in 3- & 4-inch, striking in 6- & 8-inch, & simply astonishing in 10- & 12-inch! Herschel & the other classic observers apparently never noticed this effect - is it possible the planetary has physically evolved over the past 150-200 years to shift its primary emission lines into the retina-sensitive parts of its spectrum? A number of other planetaries exhibit similar behavior, but none perhaps so strikingly as NGC 6826. A delightful observing experience! D=3,300LY

Foxhead Cluster Small, faint but rich cluster of 150 suns in a great MW setting. It displays "Two V-like 'ears' pointing NW." "A hazy glow bespangled with a few dim points of lights" in small scopes. "Two intersecting strings" of stars visible in 6-inch. Grows more attractive with increasing aperture. Fairly remote for an OC: D=7,300LY

Veil/Filamentary/Network/Lacework/Cirrus Nebula "Sky's brightest (?) supernova remnant." Exploded 150,000 years ago! Largest in N sky & one of most readily seen. Two halves of a vast gas bubble nearly 3 degrees in diameter. NGC 6960 (the W segment) is easiest to find since it passes right across the naked-eye star 52 CYG. "Nebulous ray extending N & S of 52." "A difficult & indistinct object in a 6-inch glass." "One of the finest deep-sky objects! Wraith-like filaments." NGC 6992-5 (E half) is somewhat bigger & brighter but harder to find. "Ghostly arc...a thin twisting spike." A "Large nebulosity in a curve." "Like a ghostly white rainbow, over 1 degree in length." "Complex is unquestionably one of the best-known telescope sights of the summer sky...(yet) always proves challenging through...amateur telescopes." "Two wonderful lace-like gaseous nebulae, forming together the 'Bridal Veil Nebula'...the small telescope will show nothing of its true splendor." Like "The long & shelving undulations of a thin cataract of light, as it slips from star to star in its shining fall through space." "A diaphanous filigree of star-spangled red, white & blue" (as seen in photographs). "The shards of a supernova; the spawn of minds to come." A very dark night, low X, WF & aperture all essential for good view. Using a nebula filter greatly enhances the visibility of this "broken bubble," its edges looking like "ragged cotton" in a 10-inch. Both halves can be seen *together* in big binoculars & RFTs! D=2,500LY

NGC 7027 CYG (7)	21 07 +42 14	PN?	9.0/11.3	18"x11"	

! Stephan's/Webb's Proto-Planetary Small, intense, eerie-looking PN. Fuzzy blue star in 3-inch - a fascinating sight in 6-inch & larger apertures. "Planetary, like an 8.5 (mag.) star, about 4" (??) dia." "Bluish-green; protoplanetary?" "Unusual." "Strange object originally believed to be a star, then a PN & now a bizarre emission object."! "High surface brightness, with double lobe structure." Like famed nearby NGC 6826 (see above), it too 'blinks'! "Has richest spectrum of all the planetaries." D=3,000LY

Gamma DEL (20)	20 47 +16 07	DS	4.5, 5.5 K1IV, A2I	10"	

! Beautiful yellow & pale-green pair with the tiny **Ghost Double** Struve 2725 (7.6, 8.4, 6") in field, 15' to SW. A "showpiece DS." "Golden yellow, greenish blue." "Golden, blue-green." "Yellow & turquoise; a charming pair." "Yellow & light emerald; beautiful." "Both yellowish, the fainter star often appearing slightly tinged with green." "Rich duo - gold & green." "Magnificent." "Reddish yellow, greyish lilac."!? Lovely sight in a 3-inch at 45X - in a 6-inch or bigger, one of the very finest of its class. Yet, the hues of bright pairs are best seen in small glasses. In a 2-inch at 100X, the stars' big beautifully-tinted diffraction disks & rings appear nearly touching one another! Only very slight orbital motion seen, but the two suns are gravitationally bound as a CPM system. Although both doubles are at the same distance (D=100LY), they're moving through space in nearly opposite directions! "A splendid sight."

NGC 6905 DEL (6)	20 22 +20 07	PN	11.9/13.5	44"x38"	

Blue Flash Nebula An unusual & overlooked PN, visible in 5-inch & a fascinating sight in 10-inch or larger scopes. "Small, faint, misty, ill-defined, closely surrounded by several faint stars." "Fine though small...in a course cluster." An "Oblong bluish blob; soft & small." "A beautiful ring-shaped nebula with faint central star. Larger instruments show it as one of the finest planetaries in the Summer sky." Lies near the DEL-SGE border in a rich MW field. Bluish tint obvious in a 7-inch. D=4,200LY

Mu DRA (11)	17 05 +54 28	DS	5.7, 5.7 F7V, F7V	2"	

! Cozy, identical-twin binary with 480-yr. period. "A beautiful miniature of Castor (in GEM)." "Neat...both white...considered a miniature of Castor, but the stars are too nearly equal to bear out the resemblance accurately." "Tiny pair; good test for smaller scopes." "A pretty pair." "Yellowish orange & bluish white, very colorful, very close, almost touching." Was closest at 1.9" in 1988 - now slowly opening to a maximum of 4.5" in 2180. A 5-inch just separates them at 70X on a steady night. Sweet! D=82LY

Nu DRA (13)	17 32 +55 11	DS	4.9, 4.9 A5, A5	62"	! Another perfectly matched pair - but brighter & much wider than Mu. Easily split in binoculars & a lovely sight in a 2-inch glass at 25X. "Beautiful fieldglass double." "A grand object" in 4-inch, "exceptionally fine" in 10-inch. "Both yellow-white." "Wide white twins." "Dramatic." "Regarded as one of the finest binocular pairs." Nu is also one of the few wide pairs that are still attractive as seen in larger instruments. A CPM system, one of these stellar twins is an unresolved spectroscopic binary. D=120LY
Epsilon DRA (6)	19 48 +70 16	DS	3.8, 7.4 G8III, F6	3"	Tight magnitude-contrast duo for medium apertures. "Yellow, blue. Contrast very pleasing." "Light yellow, blue, elegant." "A striking object." "Slight color contrast." Some early observers had trouble seeing the companion & suspected it was variable??
Psi DRA (9)	17 42 +72 09	DS	4.9, 6.1 F5IV, G0V	30"	! Easy wide combo for small glasses. "Yellow & lilac." "Neat...both pearly white." "Yellow & orangish yellow, nice separation & brightness." Pretty sight! CPM pair.
17-16 DRA (11)	16 36 +52 55	DS	5.4, 6.4, 5.5 B9V, A1V, B9V	3", 90"	! Nice triple like Mu BOO, but primary has companion & the components are more equal in brightness. "A striking triple system." "Pale yellow, faint lilac & white." "All white in excellent field - close & wide combination." Neatly split in a 4-inch at 45X, just perfect in 6-inch at 50X! Close couple likely a binary, the third star CPM.
39 DRA (6)	18 24 +58 48	DS	5.0, 8.0, 7.4 A3V, F6V, F8	4", 89"	A big-scope trio. "Wide, impressive, colorful!" in 10-inch. "Yellowish white, bluish, ash; (& again) white, ruddy, lilac." "Impressive triple." Close pair needs at least good 5-inch for clean split & 8-inch or more to appreciate the delicate beauty of this object.
41-40 DRA (9)	18 00 +80 00	DS	5.7, 6.1, 7.5 F7, F7	19", 222"	Neat easy twins with fainter third star some distance away in field. "A beautiful pair." "Yellow, paler yellow...grouped finely with a smaller lilac star." "White & off-white ...nice pair in nice field (with) another star." "Fine CPM pair." Easy in 2-inch glass.
RY DRA (0)	12 56 +66 00	SS	6.8-7.3 N7.7 (C3)	---	Ruddy carbon star - one of reddest in *Sky Cat 2000.0*. "Deep red." Tint obvious in a 3-inch & nice sight in 6-inch - a glowing stellar ruby! A huge pulsating, semiregular variable sun, like most members of its colorful & restless class. Period 172 days.
UX DRA (1)	19 22 +76 34	SS	5.9-7.1 N0 (C6)	---	Another stunning red carbon star with 168-day period of variability, easily visible in small glasses. These crimson jewels are great objects to show (& surprise!) those who think that all stars are colorless - only "white" - as many believe. "A striking sight."
NGC 6543 DRA (17)	17 59 +66 38	PN	8.8/10.5?	22"x16"	! **Cat's Eye/Snail Nebula** Very bright blue-green egg with nuclear sun! "Luminous blue...magnification discloses star center." "Very curious...very luminous disc, much like a considerable star out of focus." "Blue spheroid - like a snail in photos." "A

remarkable object." "Brilliant small PN, cobalt-blue-colour; stellar nucleus, flashing light, very singular." "Bluish ellipse." "Remarkably bright, pale blue...easily found, & always above the horizon...an object of much interest." "Disc with very ill-defined edges." One of brightest of its class, easily seen in 3-inch. Striking in 8-inch & a truly magnificent sight in 14-inch - showing "Swirling complexity...greenish cast...(&) a dark circular void...directly adjacent to the central star." The nebula has been rated as bright as magnitude 8.1 & the central star itself at 9.5 visually, but the latter is "not easy to distinguish in the bright mass of nebulosity." The nebula's distinctive color is "strikingly vivid in large scopes." "One of the most glorious in the sky." D=3,500LY

Name	Coords	Type	Mag	Sep	Description
NGC 5907 DRA (4)	15 16 +56 19	SG	10.4	12'x2'	**Splinter Galaxy** A well-named but seldom-observed wonder. "Long thin streak." "Extremely large, narrow streak." "Fine edge-on with dust lane." A "silver sliver." "Grand spiral, one of the finest examples of its class. Large, beautiful, surpassed as an edge-on spiral only by NGC 4565 (COM)." Rather dim - needs dark night & at least 8-inch to appreciate. "Magnificent." Other galaxies lie nearby. D=35,000,000LY
Epsilon EQU (10)	20 59 +04 18	DS	6.0, 6.3, 7.4 F5III G0V	0.8", 11"	"A most delicate triple star." "Yellowish, yellowish, ashy white." Wide pair easy in small glass; close one varies from an impossible 0.1" (1920) to a resolvable 1.1" in a 101-year period. Binary currently needs at least a 6-inch for definite split. D=200LY
Lambda=2 EQU (8)	21 02 +07 11	DS	7.4, 7.4 F8	3"	"Beautiful pair." "Elegant." "Striking." "Exceptional." "Like a pair of bright, foreboding eyes!" "Very neat...lovely object." Both "very white." Easy in 4-inch at 90X.
Theta ERI (12)	02 58 -40 18	DS	3.4, 4.5 A4III, A1V	8"	**! Acamar** Radiant & easy, far-S white duo - readily resolved in 2-inch glass & a truly stunning sight in all size telescopes! "Striking pair of blue-white stars." "One of the gems of the southern sky." "Dazzling pair." Very low DEC. CPM system. D=93LY
Omicron-2=40 ERI (8) Ob. HwBK 2004 p.269	04 15 -07 39	DS	4.4, 9.5, 11.2 K1V, DA, M4	83", 8"	A triple sun involving a faint but remarkable red dwarf-white dwarf duo! Star B is the easiest white dwarf in the sky to see in small telescopes, being "the classic example." "Unusual & nearby system." "Group apparently concrete" - a CPM trio. The close B-C pair is a binary with 248-year period, whose separation ranges from about 2" to a maximum of 9" (last reached in 1990). A 3-inch will split them in good seeing, but it takes at least a 6- to 8-inch to clearly see their contrasting white & red tints. Together they offer a unique opportunity to view the two most common types of stars among the stellar population - red & white dwarfs (C & B, respectively)! D=16LY
32 ERI (17)	03 54 -02 57	DS	4.8, 6.1 G8III, A2V	7"	**!** Vividly-tinted yellow & blue-green pair - lovely in anything from a 2-inch glass to a 14-incher! "Topaz, bright green...colors 'magnifici, superbi'." "Topaz & marine blue." "Beautiful." "Superb object." "Emerald & topaz gems - one of the redeeming features of this somber region." "Yellow & aqua." "Topaz yellow, sea-green; colours in brilliant contrast." "Magnificent colors." In rather isolated region, but its "beauty... makes it well worth the search." Striking in a 6-inch at 50X. CPM pair. D=300LY

NGC 1535 ERI (16)	04 14 -12 44	PN	9.4/12?	20"x17"	**!** Pale blue-green planetary with faint central star. Somewhat difficult in 3-inch scope; 6-inch easily shows nebula but not star, 13-inch shows both. "Bright...with blue-grey disk." "Pale bluish disk with 11.5-mag. central star." "Bright & round, with low powers...but not bearing magnifying." "The most interesting & extraordinary object of the kind I have ever seen: an 11 magnitude star standing in the centre of a circular nebula, itself placed centrally upon a larger & fainter circle of hazy light." "Splendid though not very conspicuous object, greyish white colour." "Somewhat like a star out of focus, with a planetary aspect." "Like all planetaries, just a hazy dot in the sky."?? "Distinct bluish discus!" "Strong, unmistakable bluish green color." "Bright...unique ...well worth the search." "A celestial jellyfish." "One of season's finest...surprisingly large for a planetary & glows with characteristic bluish tint." "Like another Neptune."
NGC 1316 FOR (2)	03 23 -37 12	SG	8.8	7'x6' (Please see Page 95)	**Fornax A** Leading member of the Fornax Galaxy Cluster, which hosts some 18 bright galaxies & many fainter ones (9 of them within a 1-degree field centered on RA 03.5 & DEC -35.5!) at the FOR-ERI border. "Bright with intense core." "Proves to be an amazing sight when viewed through a medium or large backyard telescope & is one of the true unsung marvels of the early Winter sky." The 11th-mag. SG, NGC 1317, lies just 6' N. NGC 1316 itself is a radio source, thus its name. D=55,000,000LY
NGC 1360 FOR (2)	03 33 -25 51	PN	9.4/11.4	6'x4'	**!** "Peculiar nebula...usually classed as a planetary...uncertain." "Impressive, elongated ...central star visible." "Seems out of place in the midst of the area's many galaxies. Known to few amateurs, even though it puts on a fine show in telescopes as small as 6-inches...a bright oval disk of greyish light highlighted by...central star." "Should be added to every winter star party observing list." "One of brightest PNs in the S sky. Visible in 2-inch telescope as bluish haze surrounding 11th-mag. central star. Remarkably high surface brightness." Nucleus rated as bright as 9th-mag. by some? "An often overlooked jewel...has immediate 'wow' appeal." "A fantastic...planetary." D=980LY
Alpha GEM (21)	07 35 +31 53	DS	1.9, 2.9, 8.8 A1V, A2V, K6	4", 72"	**! Castor** One of the famed "Twin Stars." Radiant blue-white snug binary with distant orange-hued third star - a red dwarf eclipsing system known as YY GEM (9.1-9.6, & period just 20 hours!). All three easily seen in 3-inch glass. "The largest (brightest) & finest of all the DSs in our hemisphere." "The certainty of its motion fully convinced William Herschel of the existence of binary systems." "That gem of DSs...both greenish-white...a beautiful object." "Very luminous pair; one of the finest doubles." Strangely, at least two observers from the past have called Castor a "Beautiful orange-blue pair."?? A-B orbital period is around 500 years; the stars are now opening from a minimum of 1.8" in 1963 (at which time they could be seen merging over a period of just *months* in a 2.4-inch!) to a maximum of nearly 8" in 2120. Both stars are also spectroscopic binaries (with periods of 9 & 3 days), so Castor is actually an amazing six-sun sextuple system! "Astounding multiple star." "One of the most remarkable examples...in the heavens." For a thrilling view of the A-B pair's diffraction disks, use at least 150X on a 2- to 3-inch! "One of the great show objects of the sky." D=52LY

Beta GEM (2)	07 45 +28 01	SS	1.14 K0III	---

Pollux Although a first-magnitude star & much brighter than Castor, this stellar gem is largely ignored because of its more famous Twin. Yet, it's well worth comparing its orangish hue against Castor's - "Shining with a diamond whiteness in contrast to the bright golden tint of Pollux...a sparkling pair" as seen in binoculars. Pollux itself has several wide faint optical companions visible in the telescope. D=34LY

Delta GEM (12)	07 20 +21 59	DS	3.5, 8.2 F0IV, K3V	6"

! Yellow & reddish-purple duo like tight Eta CAS. "Creamy white with orange dwarf companion; brightness contrast makes pair difficult in telescopes below about 3-inches aperture." "Yellow & magenta." "Yellow & pale blue." Needs at least 6-inch to truly appreciate color contrast - striking in 12-inch! A 1200-year-period binary. D=53LY

Kappa GEM (9)	07 44 +24 24	DS	3.6, 8.1 G8III	7"

"Very delicate & beautiful." "Yellow, ashen. Also called orange & blue." "Elegant object." "Impressive region!" Some lists give the companion as faint as mag. 9.4 & at least one early observer thought it was possibly "shining by a reflected light" from the primary star - so faint did it appear to him! Perhaps variable?? Best in 8-inch & up.

20 GEM (6)	06 32 +17 47	DS	6.3. 6.9 F8III	20"

Neat easy pair for small scopes. "Yellowish white, bluish white." "Topaz yellow, cerulean blue." In fine MW field. Pretty sight in a 3-inch glass at 30X! D=450LY

38 GEM (7)	06 55 +13 11	DS	4.7, 7.7 F0V, G4V	7"

Third of three similar-looking doubles in GEM. "Attractive pair." "Yellowish, bluish or yellow white, purple." "Neat...very fine...light yellow, purple; colours so marked, that they cannot be entirely imputed to the illusory effect of contrast." "Companion changes in brightness & color; primary golden yellow, companion changes from green to blue, purple & red."!? A long-period (3,200-year) binary. Nice in 5-inch at 70X.

U GEM (1)	07 55 +22 00	SS	8.2-14.9 M4+WD	---

* Prototype of the cataclysmic variable stars - a red giant sun with blue dwarf in near-contact orbit around it! "One of the most mysterious variables." "An enigmatic star!" "Constantly watched for sharp & thrilling changes." "A remarkable dwarfish variable ...nova-like eruptive star...or miniature nova...normally a 14th-mag. object, but at intervals of several months it undergoes sudden outbursts. The complete rise from 14th- to 9th-mag. usually takes no more than two days...often accomplished in a mere 24 hours...is thus an exciting object to study. The hourly increase in light during a rise to maximum is a fascinating sight to watch." A great star to monitor but needs at least a 10-inch to see at minimum light - a 2-inch easily shows it at peak brightness!

M35/NGC 2158 GEM (21/3)	06 09 +24 20	OC OC?	5.1 11?	28' 5'

! **Lassell's Delight** A big bright, splashy stellar jewel-box with faint, tiny & remote "semi-globular" lying on outskirts, 1/2 degree SW. "A marvelously striking object. No one can see it for the first time without an exclamation of delight. Nothing but a sight of the object itself can convey an adequate idea of its exquisite beauty." "One of the finest clusters in the heavens." "A splendid sight." "Beautiful & extensive region of small stars, a nebula to the naked eye...the stars form curves, often commencing with a larger one. There is an elegant festoon near centre, starting with a reddish star."

"Presents a gorgeous field of stars...with the centre of the mass less rich than the rest... somewhat reminds one of the bursting of a sky-rocket." A "Magnificent...glorious profusion of stars...curved lines (of stars) appear like exploding fireworks...an amazing mass of sparkling stars...superb object...the grandest OC of the Winter MW." "Most of the brightest cluster stars are blue-white, though a few are yellow & orange giants. Many appear to form graceful arcs & curves threading their way throughout the cluster ...a curious absence of stars near the center." "A 'hole' in center." "Large gap across center; appears tri-partite." The "Curving rows of bright stars give an impression of... glittering lamps on a chain." "A splendid specimen." "Superior to the Perseus clusters (the Double Cluster)." "Strikingly beautiful." "Somewhat too large for normal fields." "Marvelous sight in binoculars...a dense swarm of stardust peppered with a half dozen stellar pinpoints." "The Gem of Gemini...one of grandest objects in sky...too beautiful to describe." NGC 2158 located "Just beyond a group of outliers... (appearing as) a faint dim cloud of very minute stars." "May be a transition cluster between the galactic & globular types." It certainly looks globular in the eyepiece - & also on most photos! "Globular cluster in fine field." "Mysterious little cluster." Lies near the galactic anti-center in an outer spiral arm at the rim of our Galaxy "Shining like a lighthouse beacon at the edge of the blackness of intergalactic space." It can be glimpsed in a 3-inch scope at 30X on dark nights as a gentle glow, but has been called difficult in a 4-inch. It's definitely there in a 5-inch at 50X! M35 itself hosts some 200 stars, while NGC 2158 packs at least 150 of them into its spheroidal shape. These swarms actually lie quite far apart in space: D=2,700LY & 16,000LY, respectively!

NGC 2392 GEM (18)	07 29 +20 55	PN	8.3/10.5	20"

! Eskimo/Clown Face Nebula Vivid blue PN with bright central star, visible in the smallest of glasses. "Double shell around 10.5-mag. star." "Blue-green disk." "A very remarkable phenomenon." "I found such a conspicuous nebulosity that I thought it was either damp on the eye lens or a telescopic comet." "Nebulous unit...worth many returns. A gem-field!" "Tiny vibrant blue disk." "Beautifully blue." "A ghostly bluish-green color becomes evident with larger telescopes...an impressive sight." "Curious & remarkable." "A star enveloped in an atmosphere." High X on a 12- to 14-inch reveals fascinating structure within the nebula, including "Overlapping bright rings & dark patches." "A marvelous object - a star surrounded by a small circular nebula, in which, close to the star, is a little black spot. This nebula is encompassed, first by a dark then by a luminous ring, very bright, & always flickering, owing no doubt to the unsteadiness of the atmosphere." Some sources give its size as big as 40" of arc, which it seems to live up to in larger telescopes. "Less than 2,000 years old - one of the youngest PNs." Don't miss it! Diameter 0.5 LY - D=3,000LY

Alpha HER (21)	17 15 +14 23	DS	3.1-3.9, 5.4 5" M5II, G5III+F2V

! Rasalgethi Bright orange & blue-green pair - tints very intense! "Excellent color contrast." "An outstanding DS." "Colors strong & easy to see without guessing." "Orange & emerald." "Very yellow, intense blue." "Orange & aqua." "Beautiful." "One of the nicest color contrasts in the sky - red & greenish-blue." "Lovely object,

....
....
....
....
....
....
....
....
....
....

one of the finest in the heavens." "A splendid double...the components being orange & emerald green...(I) can never forget the feeling of amazement with which (I) gazed at this celestial gem...nor the exclamation of delight when glimpsed (by a fellow observer)...one of the red letter day experiences, indelibly recorded in (my) memory." The primary is a huge pulsating supergiant, semiregular variable & one of the largest stars known - 400 to 600 x Sun's diameter! "Since the large star is variable, the pair will be found easier to divide at some times than at others" - this due to the changing magnitude contrast between the two stars! Superb in a 3-inch at 100X - awesome in a 13-inch at 150X! Orbital period of this vast system is some 3,600 years & the small star is a spectroscopic binary (note dual spectral type) with 52-day period. D=630LY

| Delta HER | 17 15 +24 50 | DS | 3.1, 8.7 | 14" |
| (14) .... | | | A3IV, G4 | |

! Delicate magnitude & color contrast optical pair - white & purple. "A study in color." "Green, ashy white; (& again) pale yellow, bluish green." "Sometimes yellow, usually white, & azure." "Pale yellow, ruddy purple." "White & violet." "Greenish white, grape red." "Green, ashen." "Green & purple." "Clear blue & violet." "Blue, white; sweet but illusory!" Separation was 25" in 1835, closed to within 9" in 1962 & is now widening again as the two stars move their separate ways nearly at right angles to each other. Best in 8-inch or larger instrument - lovely sight in 12-inch at 100X! D=94LY

....
....
....
....
....
....

| Zeta HER | 16 41 +31 36 | DS | 2.9, 5.5 | 0.7" |
| (7) .... | | | F9IV, G7V | |

Herschel's Rapid Binary Yellow & red ultra-tight pair. One of the brightest, nearest & fastest-moving visual binaries for larger amateur scopes: period just 34 years! It has now made over *six* orbits since its discovery by Sir William Herschel in 1782! "Close binary star, yellowish white, orange tint...wondrous object...baffled all endeavors (of early observers) to divide, or even elongate it (at its closest approaches)...astounding velocity...what a motion!" Requires good 4-inch to split at widest separation of 1.6" (last in 1991); the pair is now closing to another minimum of 0.5" in 2001. D=30LY

....
....
....
....
....

| Kappa HER | 16 08 +17 03 | DS | 5.3, 6.5 | 28" |
| (8) .... | | | G8III, K1III | |

! An easy, colorful yellow & orange optical pair for small glasses. "Light yellow, pale tawny or garnet." "Yellow & bronze." "Yellow & red." "Pale yellow, dull orange."

| Rho HER | 17 24 +37 09 | DS | 4.6, 5.6 | 4" |
| (15) .... | | | A0V, B9III | |

! Attractive, close, roughly-matched combo - both stars white with tinges of blue & green. "Bright white pair." "Greenish white, greenish; (& again) white, bluish." "Both green." "Beautiful, bluish white, pale emerald." Stunning in 5-inch at 70X!

| 95 HER | 18 02 +21 36 | DS | 5.0, 5.1 | 6" |
| (16) .... | | | A5III, G8III | |

! Perfect twins with unusual pale-red & pale-green tints or "Apple green, cherry red."! These amazing hues subtle but persistent in all apertures. Also called "green-yellow, red-yellow", "beryl & sardonyx", "gold-yellow & azure", & even "gold & silver"?? "Strange discrepancies as to colour." "Thus, the magnificent tints of orange & green (seen in the mid-1800s) were of a transitory nature." "Transitory...or illusionary? It seems quite unlikely that color changes of such rapidity could be physically real." A fascinating possibility in any case! "An extremely pretty pair, very bright, a charming little picture, variable colors." "Notorious among observers for discordant estimates of

....
....
....
....

100 HER (8)	18 08 +26 06	DS	5.9, 6.0 A3V, A3V	14"	
M13 HER (21)	16 42 +36 28	GC	5.9	17'	

color." "One of the most stunning, most striking color contrasts of any pair in the sky." "Beautiful object." Perhaps we should dub this the "Apple & Cherry Double" since these are the tints most observers actually see here once they've been pointed out to them! Superb sight in a 6-inch at 50X! Relatively fixed - a CPM pair. D=400LY

Another matched but wider duo, pretty even in a 2-inch. "Striking." "Superb equal pair." "Exquisite." "A beauty." "Greenish white, bluish white." "White, flushed white." "Both pale white." "Easy CPM pair." Seldom looked at - what a pity!

! Hercules Cluster The Great Globular Star Cluster in Hercules, one of the grandest sights in the heavens! Finest GC in N sky - a magnificent object! Incredible stellar beehive of up to a *million* suns! "Superb GC...finest of its class...spangled with glittering points...becomes a superb object in large telescopes." "The mere aspect of this stupendous aggregation is indeed enough to make the mind shrink with a sense of the insignificance of our little world." "Spectacular showpiece - resolved nicely at 100X (in 8-inch), but higher power will blow you away!" "Along with the Orion Nebula, is perhaps the most observed deep-sky object." "A grand GC...centrally resolved in a 6-inch telescope." "Not the biggest. Not the brightest. Just the most famous. Star chains suggest 3-D effect. Breathtaking in large backyard telescopes." "An amazing sight; the outer stars are resolved in a small telescope, & with a larger glass the entrancing beauty of the great globe of stars is revealed." "Most beautiful...exceedingly compressed in the middle, & very rich." "A showpiece...great swarm, 500 times as many stars at center than in solar neighborhood!" "Small scopes resolve individual stars throughout the cluster, giving a mottled, sparkling effect." "In a small glass a curious olivine blur, often resolvable; but viewed with apertures of increasing light-gathering power, a blazing panoply of suns of varied tints & staggering brilliancy... majestic prodigy." "Perhaps no one ever saw it for the first time...without uttering a shout of wonder." "Premier object of its class in N sky." "In larger scopes, a never-to-be-forgotten sight." "This superb object blazes up in the centre...indeed truly glorious, & enlarges on the eye by studious gazing...brilliant...an extensive & magnificent mass of stars...densely compacted." "In the greatest telescopes...an incredibly wonderful sight; the vast swarm of thousands of glittering stars, when seen for the first time or the hundredth, is an absolutely amazing spectacle." (The view of M13 in a 30-inch refractor at 600X leaves you totally breathless & in reverent awe!!!) "A mass of glittering starlight...the glamour of the light from these intermingled stars remains for the fortunate observer who has the privilege of actually *seeing* them." "A Treasure-Trove of 50,000 Stars." Yet Messier was sure this object contained *no* stars! "The appearance of the heavens from a point within the HER Cluster would be a spectacle of incomparable splendor; the heavens would be filled with uncountable numbers of blazing stars which would dwarf our own Sirius & Canopus to insignificance. Many thousands of stars ranging in brilliance between Venus & the full Moon would be continually visible, so that there would be no real night at all." Such is the setting for the classic sci-fi short-story *Nightfall*! "Three dark 'lanes' or rifts in its interior, beautifully seen" forming a 'Y' SE of the core. "Hazy mothball" at low X. "Splendid." Using averted vision, let M13 drift through the eyepiece field to reveal numerous faint outlying stars in "hairy-looking curvilinear branches" - star chains radiating from the globe's edges. "Streamers curving outward as though wafted by a celestial breeze."! Its diameter is at least 160LY. D=24,000LY The faint 12th-mag. SG, NGC 6207, lies 1/2 degree to NE. "Sweet & easy!" in 8-inch. "Offers one of sky's best examples of extreme depth of field - about 2,000 times the distance of M13" itself! But this tiny remote spiral isn't an easy object to see under light-polluted skies - even in a 6-inch.

M92 HER (16)	17 17 +43 08	GC	6.5	11'	

(Please see Pages 95-96)

! Overshadowed Globular Beautiful bright swarm with very intense core, seemingly even richer than M13 centrally. Yet it's "One of the unsung glories of the heavens, as observers frequently ignore this GC in favor of its famous neighbor...an unappreciated treasure." "Very fine cluster, though not equal to M13; less resolvable; intensely bright in centre." "Stars brighter & more compressed than in M13, but blaze resolved by glimpses" in 9-inch. "Magnificent GC of small stars." "Fine...with luminous center." "Rival of M13...both are brain-burners!" "Only slightly inferior to its more famous neighbor M13 which overshadows it." "Beautiful...with streaming starlanes." "Looks only half as large (as M13) - a grand object...many stars seen in bright central region. Surrounding is a fainter glow that is also star-studded." "A beautiful rich GC which in almost any other constellation would be considered a major show object. The view in large instruments is stunning beyond words; the countless star images run together into a dazzling central blaze...equaled by only a few of the globulars." This is precisely its appearance & ranking in 12- to 14-inch aperture scopes! D=26,000LY

NGC 6210 HER (13)	16 44 +23 49	PN	9.3/12.9	20"x16"	

! Small, bright PN showing an intense featureless blue disk. Looks stellar at 30X in a 3-inch - 45X on 4-inch reveals definite disk & becomes more fascinating as aperture increases. "Very starlike blue planetary." "A blue-green ellipse." "Robin's-egg blue." "Bright...blue-green." "Small, not sharply defined, exactly like a star out of focus, bearing power well, 111X of 5.5-inch showed a glow round it." "Cobalt blue colour." "Intense blue." "Bright blue disc with crisp edges." Looks "about 8th-mag. visually." "Curious object." "A homogeneous disk...distinctive color...described variously as blue, green, aqua or turquoise...holds up well under higher magnification...best views will be enjoyed at 150X or more." Seems only about half size given. Also carries the designation Struve 5N from a list of nebulae discovered during his famed survey of the sky for new DSs. See NGC 6572 (Struve 6N) in OPH for a similar gem. D=3,600LY

Epsilon HYA (10)	08 47 +06 25	DS	3.3, 6.8 F0V+KIII, F5	3"	

Close duo for medium apertures. "Neat white pair." "Beautiful object - pale yellow, purple; (& again) silvery white, smalt blue." "Beautiful but difficult DS of contrasty colours...yellow & blue." A 5-inch at 100X splits it in good seeing. Binary with an 890-year orbit - & the primary itself is an ultra-close (average separation 0.2"), ultra-rapid (15-year period!) visual double in large observatory-class telescopes. D=150LY

54 HYA (10)	14 46 -25 27	DS	5.1, 7.1 F2III, F9	9"	Nice double for small glass. "Yellow & violet." "Red, blue." "Very beautiful...pale orange, violet tints." "Both white. View at culmination!" due to low DEC. "Easy... yellow & purple." One of "A legion of vari-tinted multiples, quite bewildering tho low in our latitudes" to be found in this long, winding constellation. A CPM system.
N HYA=17 CRT (6)	11 32 -29 16	DS	5.8, 5.9 F5, F5	9"	A pair similar to 54 HYA but perfectly matched in brightness. Has a "split identity" - as seen by its dual designation. Before IAU constellation boundary reassignment, was in CRT but now lies well within HYA. "Neat...lucid white, violet tint." "Both yellow-white; a striking pair!" "Yellowish equal, nice color & separation." "Very fine." A pleasing sight in 3.5-inch glass at 40X & still attractive in 10-inch at 80X. CPM duo.
U HYA (3)	10 38 -13 23	SS	4.8-6.5 N2 (C7)	---	! "Very red carbon star." "Fine spectrum." "Deep red variable star" - a semiregular of 450-day period. Bright & easy - "A gem glowing fire orange" in 4-inch glass at 45X. An interstellar 'traffic light'! Sadly, red stars are overlooked as deep-sky showpieces.
M48 HYA (9)	08 14 -05 48	OC	5.8	30'	! Big bright, "very loose...sparce sprinkling" of some 50 stars spanning half a degree of sky. "A superb object in 4-inch." "Group of pretty uniform 9th-mag. stars, with a profusion of lesser ones." "Isolated, but bright & curious." "A splendid group, in a rich splashy region of stragglers." "Giant rich naked-eye OC." "Not so round - not so firm - not so fully packed - good binocular object!" "Somewhat triangular in shape - approximate size of the Moon." "Triangular in outline...dominated by a central chain-like grouping of...stars." "A splendid splash of stars." Lovely sight in 3-inch at 30X. "One of the mysterious missing Messier numbers" - lost & then found! D=1,900LY
M68 HYA (9)	12 40 -26 45	GC	8.2	12'	A far-S, neglected GC. "A beauty!" "Well resolved into glitter of 13th-mag. stars" in 8-inch. "Imposing." "Large round...very pale...mottled." "A pleasing object in adequate instruments." A 6-inch on a good night will "reveal multitudes of its stellar components, while a 10-inch glass begins to give a hint of the true splendor of the group." A 12-inch shows "Innumerable stellar points uniformly packed across an 8' area." "Beautiful." Needs good seeing. Contains at least 100,000 stars! D=45,000LY
M83 HYA (10)	13 37 -29 52	SG	8.0	11'x10'	! "Large spiral...one of the brightest galaxies...in the heavens." "Ranks as one of the 10 brightest in the sky." But also "A frequently neglected showpiece" due to its low altitude. "An excellent object despite being so close to the horizon." "Triply branched spiral...very bright, very large. Nuclear condensation near center, & the entire nebula of S-shape backward." "Large & diffuse; tough from N latitude." "Face-on...spiral arms can be traced in apertures of 6-inches." "The 4-inch revealed enough to suggest M83 must be a magnificent object for S Hemisphere observers." "Finest face-on S-type spiral in the sky." "Has shown a remarkable number of supernovae" - 4 in just a 50-year period! - an "exceptional...frequency." Needs a dark night & clear S horizon due to its low DEC - definitely worth the wait! "Clearly a delightful spiral for small 'scopes...beautiful object." One of the finest examples of its class! D=10,000,000LY

NGC 3242 HYA (19)	10 25 -18 38	PN	8.6/11.4	40"x35"	**! Jupiter's Ghost** Bright planetary with pale blue disk as big as Jupiter in apparent size! Among the brightest & easiest of its type - obvious in 2.4-inch glass. "One of the most amazing of the planetary class - elliptical, pale steely blue...with Wolf-Rayet star-nucleus framed in double rings." A "Double shell with 11th-mag. central star." "Very brilliant. In 1892 the writer observed that it had a bright star in its center. This has now (1917) disappeared."! "Suspected variable nucleus 10.3-10.9." "Bluish egg." "Remarkable." "Spectacular & long-neglected." "Pale greyish-white. From its size, equable light, & colour...resembling Jupiter; & whatever be its nature, must be of awfully enormous magnitude." "I found it bright; a little elliptical...of a steady pale blue light, bearing high powers." "A unique object; within a circular nebulosity two clusters, connected by two semicircular arches of stars, forming a sparkling ring, with one star on the hazy ground of the centre." (This observation was made at 1000X on a 9.5-inch refractor - a "beautifully defining glass [which] accomplished marvels."!) "One of the finest PNs anywhere in the sky...(a) homogeneous turquoise disk" seen in 6- to 8-inch scopes. "Annular bright ring apparent midway from center to outer edge" in an 8-inch. "All the main features may be discerned with a good 10-inch glass but the central star always seems fainter than the measured magnitude would indicate... generally true of all the planetaries, & illustrates the difficulty of observing a faint star against a background of luminous nebulosity." Astounding sight in 13-inch refractor at 190X (& simply unbelievable in a 30-inch refractor at 600X!). But detecting the central star is tough at all magnifications - here again "because of the overwhelming brightness of the nebula itself." Sometimes called the "Eye Nebula" or "CBS Nebula" for reasons that are obvious in looking at its photos! "Ghostly greenish object - 'Cat's eye' appearance like CBS-TV logo." "Striking blue." "A good sky-blue." D=3,300LY
8 LAC (9)	22 36 +39 38	DS	5.7, 6.5, 10.5, 9.3 22", 49", 82" B1V, B2V		A delicate quadruple star. "Double-double - superb!" in 10-inch. "A-B both white, C greenish, D blue." "First two white, perhaps tinged with yellow, 10.5 uncertain, 9.3 blue." For good view of all four, at least a 6-inch is needed. Some observers consider this a fine DS with two fainter companions. A 7th-mag. fifth star lies 337" distant - if included, 8 LAC could then be considered a quintuple system! The A-B components form a CPM pair - & one very remote for an easy double: D=1,900LY! "Beautiful."
NGC 7243 LAC (10)	22 15 +49 53	OC	6.4	21'	"Splashy, coarse" grouping of some 40 gems. "Massed star-jewels." "A magnificent field of stars." "Fine cluster quickly followed by beautiful field with three pairs." "A large & loose cluster of stars. A neat DS forms the apex of a telescopic triangle near the middle of the group." This object - actually a faint triple - is Struve 2890, whose components are 8th-, 8th- & 9th-mag. at 9" & 73", respectively. D=2,800LY
Alpha LEO (5)	10 08 +11 58	DS	1.4, 7.7 BV7, K1	177"	**Regulus/Indigo Star** "Little Rex." Wide magnitude-contrast duo with strangely-tinted companion - one that appears as if it's "Seemingly steeped in indigo."! "A noble sun." "Brilliant." "An attractive sight in small scopes." "Wide but very delicate...flushed white, pale purple or violet tint." "Flushed white & ultramarine." "Blue-white &

white." Companion also called "deep yellow, almost golden" & "orange-red." Neat pair in 3-inch, but small star's unusual hue more obvious (& fascinating!) in 6-inch or larger - especially pretty in big refractors! A "Dazzling...jewel." CPM pair. D=78LY

Gamma LEO (21)	10 20 +19 51	DS	2.2, 3.5 K0III, G7III	4.4"	

! Algieba Superb 600-year-period binary - two radiant golden suns with traces of red & green - easily split in a 3-inch glass at 75 to 100X. "Finest double in the N sky." "A spectacular DS." "Brilliant." "Splendid...a most beautiful object - bright orange, greenish yellow." "Two yellow translucid diamonds." "Gold, greenish red; (& again) yellow, greenish yellow; (then) yellow, deeper yellow." "A magnificent binary pair." "Orange & pale yellow." "Blazing, close unequal white duo." As with most bright close pairs "Best observed when not quite dark or in moonlight (reducing glare from the stars). Yellow & green." "One of the most beautiful pairs in the sky." "Golden yellow giants...form an exceptionally handsome double." Minimum separation of 0.4" occurred in 1741 & the pair is slowly opening to a maximum of 4.6" in 2063. Seen through a 2- or 2.4-inch glass (or in larger scopes masked down to that size) at 100X to 150X, Algieba displays two lovely golden disks hugging each other! D=90LY

54 LEO (10)	10 56 +24 45	DS	4.5, 6.3 A1V, A2V	6"

! Little-known greenish-white & bluish-white pair - a lovely object in all size scopes! "White/blue-white; nifty pair!" "Beautiful...white, grey." "Green-white, blue." "Very fine." Contrasts nicely with Algieba's golden hues! Striking in 6-inch at 50X. One observer calling this an "unequal white" pair described it as "unimpressive"?! Take a look for yourself - it's sure to become a personal favorite! CPM double. D=150LY

Iota LEO (7)	11 24 +10 32	DS	4.0, 6.7 F2IV, F4	1.6"

Tight binary with 190-year period; stars nearest at 0.6" in 1942 & are now opening to widest separation of 2.7" in 2074. "Close & difficult" in small scopes. Needs 6-inch for definite split. "Both stars are yellowish-white." "Bright orange, greenish yellow." "Amber & turquoise." "Pale yellow, light blue...a beautiful object." "Primary yellow; companion changes color from blue to yellow to indigo & purple."!? D=69LY

R LEO (5)	09 48 +11 26	SS	4.4-10.5 M7III-M9III	---

Peltier's Variable Star The object that started famed variable star observer & prolific comet discoverer Leslie Peltier (referred to as "the world's greatest nonprofessional astronomer" by Harlow Shapley) on his long career in 1918. "Appears strongly red when at maximum." "A fiery, pulsating beacon!" "Strong reddish glow...a seasonal favorite." "One of the most fiery-looking variables on our list - fiery in every stage... really a fine telescopic object in a dark sky, when its colour forms a striking contrast with the steady white light of the 6th-mag. star a little to the N." This star is 19 LEO, which lies just 10' NW of R; 18 LEO also lies in the field about 1/2 degree to NW. "One of the finest of its mysterious class." "Noted for the peculiar intensity of its red light, best described as a rosy scarlet with often a seeming touch of purple." The hue of this long-period (321-day) variable is easily seen at maximum light in binoculars but at minimum needs at least a 6-inch to enjoy. "One of the largest stars known - as big as Jupiter's orbit" when at its maximum inflation! A restless sun! D=600LY

M65/M66/NGC 3628	11 19 +13 05	SG	9.3	10'x3'		! **The Leo Triplet** of spirals, visible even in a 3-inch glass. M65: "spindly", "dark
LEO		SG	9.0	8'x4'		lane of nebulosity...girdles outer circumference", "remarkable...beautiful". M66: "oval
(10/10/5)		SG	9.5	15'x4'		disk...prominent central core...spiral-arm halo", "magnifica", "lovely", "mottled or
....						clumpy", "beautiful specimen". NGC 3628: "big edge-on with dust lane", "long, thin
....						disk...bisected by prominent dust lane". M65 & M66 together: "form a pretty pair",
....						"elongated visually in different directions", "very curious & mystifying, worth diligent
....						study", "inconceivably vast creations", "as fine a pair...as you could hope for", "form a
....						rather striking pair in large...binoculars". And all three galaxies together: "form a
....						noble group for the small telescope." M65 & M66 are 21' apart, while NGC 3628 lies
....						35' N of M66. This clan is a fascinating sight in an 8-inch at 60X! D=30,000,000LY

M95/M96/M105 LEO	10 44 +11 42	SG	9.7	7'x5'		! Another fine spiral trio in LEO - all three Messier-objects - again all lying within the
(10/10/8)		SG	9.2	7'x5'		same low-X eyepiece field! M95: "central bar", "sharp stellar nucleus...outer ring...
		SG	9.3	4'x4'		spiral-arm halo...bar that connects the outer ring with core", "a lucid white nebula,
....						round & bright", "theta structure". M96: "nebulous glow surrounding a brighter, oval
....						heart", "silver grey, intense central region", "very bright nebula". M105: "circular
....						spot", "looks like an unresolved GC", "nearly circular disk...bright stellar nucleus",
....						"little ball of fuzz", "interesting field with three...nebulae". These two additional
....						objects are NGC 3384 & NGC 3389, a 10th-mag. elliptical & a dim 12th-mag. spiral,
....						respectively. M95 & M96 together: "pair up to create another fine galactic duo", "two
....						large round nebulae", "prominent...low power pair". Of the trio's environs: "A nest
....						of nebulae! Worth many adventures." "A region where these luminous masses are
....						scattered over the vast concavity of the heavens in truly boundless profusion...mighty
....						laboratories of the universe." Both the M65/M66 & M95/M96 clans are members
....						of the Leo Galaxy Group - a "sub-cluster" of the vast Virgo swarm. D=30,000,000LY

NGC 2903 LEO	09 32 +21 30	SG	8.9	13'x7'		! One of the best galaxies missed by Messier & among the finest in the NGC. "Big,
(9)						bright & beautiful, with brighter center!" "Brightest galaxy in Leo." "Conspicuous."
....						"Isolated spiral." And part of its appeal *is* its isolation - far from the confusing crowds
....						of galaxies in LEO itself, & in VIR & COM to the E, it's easily located & identified.
....						"Elongated, faint, with double nucleus" - thus the dual designation NGC 2903-5 that's
....						sometimes given. "A bright-class white nebula. It requires...the closest attention &
....						most patient watching to make it a bicentral object." "A double nebula, each having a
....						seeming nucleus, with their apparent nebulosities running into each other." Actually
....						a dim starcloud lying just 1' N of the true nucleus itself, NGC 2905 is a difficult catch.
....						Its host galaxy "Hangs like a misty jewel." "Easy & impressive." D=30,000,000LY

Gamma LEP	05 44 -22 27	DS	3.7, 6.3	96"		! Attractive wide pair for low-X scopes & binoculars. "May be rated excellent or poor"
(9)			F6V, K2V			depending on magnification used! "Pale yellow, garnet - easy." "Light yellow, pale
....						green." "Yellow & orange." "Yellowish & faint blue." "Splendid DS awash in vivid
....						color. Primary sun is a striking yellowish orb...secondary star appears as an orangish
....						point of light...they form one of the finest...in the binocular sky." "This brilliant wide

pair makes a beautiful low-X object." "Pleasing color contrast...easy & appealing." Very pretty in 2-inch at 25X; too spread out in 8-inch at 80X. CPM duo. D=29LY

R LEP (7)	05 00 -14 48	SS	5.5-11.7 N6 (C7)	---	**Hind's Crimson Star** Mira-type long-period variable & prime example of the reddest stars known - but quite faint during much of its 427-day cycle. "A gleaming crimson jewel." "Of the most intense crimson, resembling a blood-drop on the background of the sky; as regards depth of colour, no other star visible in these latitudes could be compared with it." "Radiating a magnificent red hue...intense ruddy color when near maximum." "Deep orange star! Like a motorcycle tail-light." "Intense blood red." "Wine-red - Betelgeuse & Antares are mere 'pale shades' in comparison." "Famous." "Celebrated." "Remarkable." "Gleams with the sparkle of a brilliant stellar ruby against a velvet black backdrop of stardust." "Intense smoky red." Actually reddest when at minimum & hardest to see! "As with most other variables...increase of light brings with it a paling of colour. Near maximum, intense redness give place...to a coppery hue." A vast carbon sun! Best in 8-inch & up most of its period. D=1,500LY
Herschel 3780 = (3) NGC 2017 LEP	05 39 -17 51	DS/OC	7.5, 8.5, 8.4, 8.1, 9.5	89", 76", 129", 60"	Unusual object with "dual personality" - a multiple star that's also classified as an OC! "Beautiful cluster for small telescopes." "Scattered group = multiple star." "Small but remarkable star cluster...5 stars...2 close binaries...so actually a family of at least 7 related stars." "A beautiful multiple star." "Interesting group of 8 stars." A "Small weak clump...with colors ranging from yellow & orange to blue." Not really much of a cluster, but very neat as a multiple system! Nice starry gathering in 6-inch at 60X.
M79 LEP (10)	05 24 -24 33	GC	8.0	9'	**Winter's Lone Globular** The season's only worthwhile example of its class (most globulars hover around the galactic center in the Summer sky), although on the dim & small side. "Tolerably bright...blazing in centre; higher powers showed it mottled." "Compact but rich, partially resolved in 6-inch at 100X." "Fairly bright, milk-white." "Impressive." "Not one of the more brilliant globulars, & becomes a truly impressive object only in rather large telescopes" - which do "turn M79 into a magnificent sight." "Beautiful & extremely rich" as seen in an 18-inch. "Rich, compressed & faint (with) bright center" in 8-inch. A "mottled nebulosity" in 5-inch at 100X. Definitely needs aperture to be fully appreciated! D=50,000LY The tinted triple star 41 LEP (5.4, 6.6, 9.1, 3", 62", G0, A3, K0) lies SW of M79, within the same wide low-X eyepiece field.
Alpha LIB (5)	14 51 -16 02	DS	2.8, 5.2 A3IV, F4IV	230"	**Zubenelgenubi** A bright, wide CPM duo for low-X telescopes & binoculars. "Pale yellow, whitish. Wide." "Pale yellow, light grey - the two form a fine though wide object." Sweet in a 2-inch glass at 25X & also in a 4-inch RFT at 16X! D=65LY
Beta LIB (5)	15 14 -09 23	SS	2.6 B8V	---	*** Zubeneschamali/Emerald Star** Claimed to be the only "green" star readily visible to the unaided eye! "Inserted for its beautiful pale green hue, very unusual among conspicuous stars; deep green, like deep blue, is unknown to the naked eye." "Pale emerald unit, resembling Uranus." "Unusual green hue." "Celebrated as one of the

few bright stars to show a distinct greenish tinge." Depending on sensitivity & color response of your eye, as well as atmospheric conditions, may or may not show such a cast. Scopes in the 4- to 6-inch range seem to do best at revealing it - & especially reflectors. (Color perception becomes difficult in the presence of too much - as well as too little - light, which is why the hues of the brighter doubles & first-magnitude stars are usually best-seen in small glasses!) Defocusing the image slightly may help.

			(Please see Page 96)		
Struve 1962 LIB (6)	15 39 -08 47	DS	6.5, 6.6 F6V, F6V	12"	Matched-white duo easily split in a 2.4-inch & a pretty sight in all apertures. Perfect in 3-inch at 30X! "Commanding." "A striking object." "Very fine." CPM system.
Xi LUP (4)	15 57 -33 58	DS	5.3, 5.8 A3V, B9V	10"	"Neat pair of...white stars well seen in small telescopes." "Splendid." "Very fine." Despite its rather low altitude, well worth searching out. A fixed system. D=120LY Lupus also hosts many other fine pairs (including Epsilon, Pi, Mu & Eta), & other deep-sky wonders - most of them quite low in the sky as seen from mid-N latitudes.
12 LYN (13)	06 46 +59 27	DS	5.4, 6.0, 7.3 A3V	1.7", 9"	! Close triple for medium apertures. "Very small trio; all tight & white." "Beautiful." "Gorgeous." "Fascinating." "Superb." "Two greenish white, (one) bluish; (& again) two yellow white, (one) ruddy." A "Delicate...curious object - white, ruddy, bluish." Close pair a binary with 700-year orbit & "Fine colors...bluish white & orangish red." A three-sun CPM system. Seldom observed. Best in 6-inch or larger glass. D=140LY
19 LYN (6)	07 23 +55 17	DS	5.6, 6.5, 8.9 B8V, A0V	15", 215"	"Attractive DS for small telescopes." "Bluish white & orangish, contrasting colors with good separation." "White & lilac." "Both white." Also a "Course triple star... A...white, B & C both...plum colored." And counting both the 9th-mag. at 215" & an 11th-mag. one at 74", a "Fine quadruple star, a beautiful sight." A-B easy in 2.4-inch.
38 LYN (9)	09 19 +36 48	DS	3.9, 6.6 A1V, A4V	3"	Bright tight pair, well-resolved in 5-inch at 100X. "Neatly double, close...beautiful & delicate...silvery white & lilac." "Greenish white, blue; (also) yellowish white, tawny." Tints subtle & elusive. "Vivid stellar horseshoe" formed with surrounding field stars.
NGC 2419 LYN (2)	07 38 +38 53	GC	10.4	4'	* **Intergalactic Wanderer** The most distant globular known (except for those in other galaxies): D=300,000LY! Yet clearly visible as small, faint ball of light near two stars in 4-inch at 45X on dark night. "Difficult to pinpoint - an undefinable glow." Its great distance from our MW "Suggests the possibility that NGC 2419 is an independent *intergalactic* object" - wandering between the galaxies of the Local Group! Here truly is "Something to ponder when you gaze its way." Sparkles in a 14-inch at 200X.
NGC 2683 LYN (2)	08 53 +33 25	SG	9.7	9'x2'	"Nearly edge-on spiral, very bright." "Very beautiful object." A "Bright oval...with splendid centre." "Large, fairly bright...spindle." It's "One of the brighter springtime galaxies." Yet, "Unknown to many deep-sky enthusiasts." "Scarcely worth the search with 3.7 (inch), but in a very fine district." "Nebulous mass. Quite interesting" seen in 10-inch. "Distinctive cigar shape...not in Webb or Smyth yet easy object for 4-inch."

Alpha LYR (10)	18 37 +38 47	DS	0.0, 9.5, 9.5 A0V	63", 118"

! Vega The "Harp Star", 'Summer Triangle' member & former Pole Star! Has several faint optical companions making it a multiple star, but the real attraction is its radiant self. "Ice Blue Vega." "Resembles an old-mine Brazilian brilliant of purest water intensified to infinity!" "Arc-light of the sky." "Use this brilliant beacon...to clean out your head!" "Beautiful pale sapphire." "Dazzling." "Offers a glorious blaze." "Its colour to me pale sapphire - a lovely gem." "Distinctly bluer than Sirius." The closer companion has been described as bluish, smalt blue & also orange by observers. Vega is "Celebrated" as the very first star ever photographed, in 1850. And it will become the Pole Star again in 14,000AD, at which time it will be less than 5 degrees from the N celestial pole & "Gazed at as the polar gem" of our skies! Vega also served as the source of the first extraterrestrial artificial radio signal received in Carl Sagan's sci-fi book & movie *Contact*. Lovely in any size glass - but in large instruments, a dazzling blue-white diamond! In traditional big refractors having a slight "blue excess" to their color correction, this great star appears nearly *pure blue* with a lovely tinge of purple - an amazing & never-to-be-forgotten sight! Lyra is truly "Adorned by one of the great leaders of the firmament." Its attendants can be glimpsed in a 3-inch glass. D=26LY

Beta LYR (11)	18 50 +33 22	DS	3.3-4.3, 8.6, 9.9, 9.9 B7V+A8	46", 67", 86"

Struve's Eclipsing Binary A striking multiple star - an amazing eclipsing binary (period 12.9 days) set within a faint starry triangle! The famed 20th-century astrophysicist Otto Struve (grandson of the great DS discoverer) spent most of his career studying Beta's many mysteries. Its variability is easily visible to the unaided eye; as with Algol (Beta PER) & other members of this class, we're actually witnessing the mutual revolution of two giant suns (here nearly touching each other!) far out in the depths of space with nothing more than the human eye! For the small telescope, "A delicate quadruple system; (A) very white & splendid, (B) pale grey, (C) faint yellow, (D) light lilac." The A-B pair has been called "creme & blue" & "yellow & white." Also, "Off-white & blue, very pretty with wide separation & brightness difference." "Truly a star for all observers." A 4-inch glass at 45X shows it all! D=860LY

Epsilon-1/Epsilon-2 LYR (19)	18 44 +39 40	DS	5.0, 6.1, B2V, F1V	2.6"
		DS	5.2, 5.5 A8V, F0V	2.3"

! Double-Double Star Perhaps the finest multiple star in the sky! Two close binaries slowly orbiting each other at a distance of 208" apart. A good 2.4-inch glass at 75X to 100X will split them on a steady night, but larger apertures show them more vividly. The position angles (direction of companion from primary on sky) of the close pairs are nearly at right angles to each other, adding greatly to their charm. "The naked eye sees an irregular-looking star near Vega, which separates into two pretty wide ones under the slightest optical aid (as with binoculars). Each of these two will be found to be a fine binary pair." "Renowned...a marvelous object." "All blue-white; gorgeous sight." "Spectacular." "Celebrated...quadruple stars are rare, & this is the finest of them." "Very elegant object...they present a vast field for contemplation." All four stars look white, with bluish-white & greenish-white tints also being reported. It's "One complex inter-related system." Orbital period of the wider duet is about 1,200 years & of the closer one around 600 years, while their mutual revolution about each other is estimated to be on the order of 10,000 centuries! This lovely "pair of pairs" is moving through space together as one vast CPM system. Epsilon-1/Epsilon-2 itself is considered to be the "Best naked-eye test in the heavens for visual acuity." D=180LY

Delta LYR (3)	18 54 +36 58	DS/	4.5, 5.6 M4II, B2V	630"	
		OC	3.8	20'	

Delta Lyrae Cluster An ultra-wide but lovely, deep reddish-orange & blue double that's involved in a sparse & rather dim star cluster (known as Stephenson-1) with 15 or so members in it. A "Beautiful color contrast of bluish-white & ruddy-orange. One of the finest DSs for small telescopes." These suns likely form a true physical pair despite their extreme angular separation & are roughly as far away as the surrounding cluster stars themselves (D=800LY). A superb binocular duo. Best at very low X in the telescope - truly stunning in a 4-inch RFT at 16X! Forms a "Glorious field."

Zeta LYR (11)	18 45 +37 36	DS	4.3, 5.9 A0, F0IV	44"	

! A tinted, wide pair for small telescopes! "Beautiful contrast." "Topaz, greenish." "Reddish & blue-green." "Greenish white & orangish white, bright & very wide." "Yellow-green, white." "Yellow, greenish; (& again) pale yellow & pale lilac." Often mistaken for Epsilon LYR in finders - but no amount of magnification will turn this duo into a 'double-double'! Very lovely in a 2-inch glass at 25X. CPM system.

O. Struve 525 LYR (2)	18 55 +33 58	DS	6.0, 7.7 G5	45"	

Miniature Albireo Yellowish-orange & blue - wide & easy. "Beautiful miniature of Beta Cygni." "Vari-colored." Often encountered when sweeping for the nearby Ring Nebula - what a pleasant surprise! Dim but nice in a 3-inch, colors vivid in an 8-inch.

Struve 2470/	19 09 +34 41	DS	7.0, 8.4 B3	14"	
Struve 2474 LYR (2)		DS	6.8, 8.1 G1, G5	16"	

Double-Double's Double Another surprising wonder from Lyra's rich treasure trove! A fainter & much more easily resolved version of the Double-Double - & in the very same constellation! The two pairs have fairly wide, nearly identical separations & also position angles on the sky (the Double-Double's are 90 degrees to each other). "What a display of symmetry in nature." "Forming a second Double-Double!" Tints for 2470 are "white, pale blue" & for 2474 "yellowish, ashy or ruddy." The pairs are 10' (or 600") apart compared to the Double-Double's 3.5' & "They sit in a wonderful starry field." Neatly split in a 3-inch glass at 30X - beautiful sight in 6-inch at 50X. CPM.

T LYR (0)	18 32 +37 00	SS	7.7-9.6 C6 (R6)	---	

Faintish example of the ruddiest stars known - striking in 5-inch & a fantastic sight in 10-inch or larger scopes! "One of the reddest stars in the sky" & also one of reddest in *Sky Cat 2000.0* "Intense." "Very red spectral type R." Another slowly-pulsating huge carbon sun with an irregular period - like some colossal beating stellar heart!

M56 LYR (10)	19 17 +30 11	GC	8.2	7'	

! "GC in a splendid field." "Semi-globular cluster, neighborhood radiant." "Often overlooked...deserves more attention." A "Beautiful little GC." "Faintish, perhaps resolvable with 3.7 (inch)." "Just barely resolved at 200X" on an 8-inch. "Readily resolvable around edges with a good 6-inch...the central mass requires a somewhat larger aperture. The field is richly sprinkled with vast numbers of faint & distant stars." "Compact, grey glow." "An impressive object...unlike most globulars, no bright core...lies in grand...field." "By itself it's not a very striking GC, but sitting in the rich field of MW stars it takes on a special charm & beauty." Indeed, this is one of those objects that definitely "grows on you" the more often you view it! And part of the reason here is the sense of great spacial depth - a 3-Dimensionality - caused by the rich starry foreground with the cluster obviously lying far beyond it. "Overshadowed (by the Ring Nebula)...another of the unsung glories of the deep sky." D=45,000LY

M57 LYR (21)	18 54 +33 02	PN	8.8/14.8	80"x60"	

! Ring Nebula Finest & best-known PN in the sky! Easily seen in 2.4-inch at 35X. Central hole is visible in a 3-inch at 45X, a definite "smoke ring" appearance in 6-inch at 50X & a truly marvelous sight in 8-inch & larger scopes at 100X or more! (Seen in a 30-inch refractor at 600X, M57 actually looks *better* than in its best B&W photos!) "The only annular nebula accessible by common telescopes, fortunately easily found 1/3rd of the distance from Beta towards Gamma. Somewhat oval, & bears magnifying well. Its light I have often imagined fluctuating & unsteady, like that of some other PNs; an illusion arising probably from an aperture too small for the object." "Famous ...finest of the annular type. With high X, the Ring becomes a globe of luminous gas, receiving its effulgence from a central star. A cardinal unit." "Best...in the heavens." "Astral doughnut." "Remarkable object." "Looks like a big smoke-ring in the sky." "Most observed PN in the sky." "Its beauty is greatly enhanced by a large field." "The prettiest PN in the sky - a grey-green puff of smoke." "One of the curiosities of the heavens." "Finest...(&) easiest of all PNs to locate." "Cosmic smoke ring easily visible in 2-inch. Ring structure & dark central hole obvious in any instrument." "Famous...a striking object in any telescope of aperture 4 inches or more. Appears just like a smoke-ring." "Superb." "Probably most celebrated of all PNs, although it's not the easiest to see. Small scopes show noticeably elliptical misty disk...larger than the planet Jupiter." In a 4-inch "Many fine streaks (across the center) in steady seeing... subjective shimmering...grey with slightly greenish tinge...6-inch shows more green." "This wonderful object - this aureola of glory...a solid ring of light in the profundity of space...however...the interior is far from absolutely dark." "It is filled with a feeble but very evident nebulous light: like gauze stretched over a hoop." "The fainter nebulous matter which fills it was found to be irregularly distributed, having several stripes or wisps in it, & the regularity of the outline was broken by appendages branching into space." Some early observers thought they had actually resolved the Ring into many "Minute stars, glittering like silver dust"! "Classic planetary...strikingly resembles a tiny ghostly donut. The peculiar color, a soft bluish-green, becomes increasingly evident with larger apertures." Its 15th-mag. nucleus can be glimpsed in a 12-inch at high X on a night that's both dark & *steady* - may be variable. Truly "One of the most beloved PN." And: "Visible in binoculars? You bet!" Diameter=0.5LY D=1,400LY

NGC 6791 LYR (0)	19 21 +37 51	OC	9.5	16'	

Faint but very rich swarm for medium to large apertures. "An unusually rich galactic star cluster in the Lyra MW." "A glorious sprinkle of over 300 13th-mag. stars" in an 8-inch. "An exceedingly rich OC." "Large, faint but very rich OC with 300 stars; a

.... faint smear in smaller instruments." Big, dim & softly sparkling - surprises you once you realize you're looking right at it & seeing it! Like a fainter version of well-known NGC 7789 (CAS). Seen in 5-inch - not on most charts. Remote for OC: D=16,000LY!

Beta MON (17)	06 29 -07 02	DS	4.7, 5.2, 6.1 B3V, B3, B3	7", 10"	**! Herschel's Wonder Star** Finest triple sun in the sky! A superb trio! Components all bluish (& sometimes yellowish) -white, forming a slender triangle & well resolved in 3-inch glass at 60X. William "Herschel's Wonder!" "Curious treble star. One of the most beautiful sights in the heavens." "Glorious triple; one of the showpieces of the skies." "Beautiful tight triple; all blue-white; will blow you away!" "All yellow." "White, yellowish white, bluish." "Three blue components." "Prime example of a triple star for small scopes. Lovely sight. Surrounded by rich, bright starfield. One thing lacking is exotic star colors - all three brilliant white." "Three yellow-white stars in comfortable spread." "Stunning triple, one pair very close, colorful, all yellowish with tints." Components "Form a curving arc of blue-white stars." "One of winter's truly spectacular multiple stars...a brilliantly white stellar triple play." "There is no color contrast in this otherwise beautiful system; all three stars appear brilliantly white." A fascinating spectacle in 6-inch & larger apertures! CPM trio. D=150LY
Epsilon=8 MON (8)	06 24 +04 36	DS	4.5, 6.5 A5IV, F5V	13"	**!** Easy gold & blue pair in rich MW setting. "Golden yellow, lilac...glorious low-X field." "Beautiful." "Magnificent field." A very lovely sight in 3.5-inch scope at 40X.
M50 MON (14)	07 03 -08 20	OC	5.9	16'	**!** Beautiful stellar jewel-box of at least 100 gems in rich field. "Bright cluster, superb, with a red star." "Brilliant cluster, straggling...to 30'; containing a red star...in a superb neighbourhood." "Fine cluster, blood-ruby star in center." "Superb object... irregularly round & very rich mass. There are certain spots of splendor which indicate minute masses beyond the power of my telescope. Numerous outliers." "Large but fairly scattered group." "Grand cluster...the brightest members form a heart-shaped figure." "Curving arcs of stars give the perimeter a rather heart-shaped outline." A "Lovely OC. A striking red cluster star prominently set amidst an ocean of blue-white suns." "Magnificent...quickly becomes a favorite target to be revisited night after night...lovely flock...a lone ruddy beacon punctuates its center." (It should be pointed out here that the frequent presence of the often-lone red or orange star set among the blue-white members of many OCs *is a direct visual indicator of stellar evolution right before our eyes!* These ruddy suns are massive giants that have already evolved off of the "main sequence" ahead of the other stars in the cluster!) M50 is perhaps at its glittering best in an 8- to 10-inch at 50X on dark, crystal-clear nights. D=2,900LY
12 MON/NGC 2244/ NGC 2237-9/NGC 2246 (11/8)	06 32 +04 52	OC DN	4.8 ---	24' 80'x60'	**Rosette Cluster/Nebula** A complex combination of a very large (twice apparent size of Moon), faint ring-shaped nebulosity surrounding an irregular cluster of some dozen or so brighter stars (and many fainter ones) centered on the 6th-mag. yellow giant sun 12 MON. Different segments of the nebula have separate NGC numbers, since typical eyepiece fields can't encompass it all at once; in fact, it's not uncommon to be looking

right through the central "hole" without realizing it! Large aperture (at least 8-inch), short-focus/wide-field (RFT) instruments give the best overall view of the combined nebula & cluster. "The cluster is likely to be the only part of this celebrated object visible in small amateur telescopes." "Pearl cluster, mag-6 star almost involved in naked-eye sunburst." "Splendid OC." "Beautiful; visible to the naked eye. Small pair near centre." "A brilliant gathering of large & small (bright & faint) stars...the latter running in rays." "Very beautiful sight in fieldglass." "Very pretty." "A distinctive rectangular pattern. Scattered within & around the stellar rectangle are more than 90 fainter stars." The nebula complex itself "May usually be detected as a formless aura of soft light encircling the star cluster" in big binoculars & RFTs on dark, transparent nights. "Wreaths of nebulosity over 1 degree in diameter...wealth of detail deserves your time." "Challenging object to see visually - faint circular glow...shimmering with a ghostly green hue." "A wonderful object" in 6-inch RFT at 24X! D=2,600LY

15=S MON/NGC 2264 (4/6)	06 41 +09 53	DS/OC	3.9	20'

Christmas Tree Cluster A big, bright gathering of some 40 stars strikingly arranged in the shape of an upside-down Christmas tree! A "Spectacular open cluster...what a distinctive shape!" "The group forms a pattern resembling an arrowhead, about 26' in length, pointing nearly due S." Brightest member is 15=S MON, a slightly-variable O-type multiple star with many faint companions (a 7.4-mag. at 3" plus half a dozen others to 10th-mag. at varying distances - including a 7.7-th mag. at 156"). "Striking triple - very colorful." "A delicate triple star, in a magnificent stellar field...greenish, pale-grey, blue." "Diamond-shaped with faint nebulosity." "Bright star group in nebulosity." The nebulosity referred to involves the ultra-faint **Cone Nebula**, a wedge or funnel of dark gas & dust extending southward from the top of the Christmas Tree. In photographs "Wonderfully outlined against glowing nebulosity...a picture of such strangeness & splendor that it scarcely seems natural...a celestial marvel...the observer is touched by a strange sensation of having been present at the drama of creation" & seated before "The Throne of God" itself! Visually it's an extremely difficult object to glimpse except in very large telescopes under excellent sky conditions. D=2,600LY

R MON/NGC 2261 (6)	06 39 +08 44	DN	---	2'x1'

Hubble's Variable Nebula Another famous nebulosity - but in this case one that *can* be seen in small instruments! "Comet-shaped." "Small, faint nebulosity containing the remarkable variable star R MON." Pulsating from 10th- to 12th-mag., it changes the apparent shape & brightness of the nebula itself as it does - along with "moving shadows cast on the cloud by dark masses drifting near the illuminating star"! Viewed with an 8-inch glass, a "Coma-shaped cloud with apex...fanlike vertex...star at head." "A dim, almost triangular-shaped wedge of fuzzy light...visible even at its minimum in small telescopes." "A bluish color is evident in large instruments. R MON itself, however, is very often a difficult object to observe, & is frequently lost in the bright glow of the nebula." "Remarkable object." "A strange & marvelous celestial spectre." "Envelopes R MON...curious reflection nebula." Due to its very small apparent size, needs lots of magnification - which it takes well because of its unusually high surface

....

....

brightness. Has been seen in binoculars! Best in 10-inch or larger at 150X & up. First object to be photographed with the 200-inch reflector at Palomar in 1949. D=2,600LY

Rho OPH (9)	16 26 -23 27	DS	5.3, 6.0 B2IV, B2V	3"	"Striking multiple star for small telescopes." A close pair, with 7th- & 8th-mag. stars at 156" & 152", respectively, on either side of it set in the midst of a large (140'x70') faint, & all-but-invisible nebulosity known as IC 4604 - itself surrounded by huge complexes of dark nebulae! A "Region of strangeness & beauty." "On an intensely black ground, in a great blank space." The close double itself can be split in a 3-inch glass at 75X. "Pale yellow, tawny...finely grouped with two 8th-mag. stars." "Pale topaz yellow, blue; (& again) yellowish white, smalt blue...two other companions in field, the whole forming a pretty group." Sweep area slowly with an RFT! D=700LY
Lambda OPH (6)	16 31 +01 59	DS	4.2, 5.2 A2V	1"	Bright tight binary for medium-aperture scopes. "Very beautiful & close." "Yellowish white, smalt blue." "Both white." Orbital period 130 years; ranges from roughly 0.8" to 1.5" in separation. An elongated egg at 100X on 4-inch; notched in 5-inch at same.
36 OPH (14)	17 15 -26 36	DS	5.1, 5.1, 6.7 K0V, K1V, K5	5", 730"	! Perfectly-matched duo with distant third star, all yellowish-orange in hue. "Splendid pair." Both "golden yellow." "Both orange! Quite a group!" Also "yellow & red." "Pretty identical twins, yellow hue." "Pair of...orange stars comfortably split by small apertures." "Ruddy, pale yellow, greyish" given for the three. Orbit takes some 550 years to complete; stars were last closest at 4" in 1917 & are now slowly opening to a maximum separation of over 6" in 2121. All three components are CPM. D=18LY!
Omicron=39 OPH (12)	17 18 -24 17	DS	5.4, 6.9 K0II, F6IV	10"	! Striking pale orange & clear blue pair - superb in a 3-inch at 30X! "Orange & blue jewels." "Beautiful." "Red, blue." "Yellow & blue." "Very fine object." CPM duo.
61 OPH (5)	17 45 +02 35	DS	6.2, 6.6 A1V, A0V	21"	Nearly-matched wide double easily resolved in a 2-inch glass at 25X. "Very neat." "Fine pair...both silvery white." Nice even in 8-inch (not too wide). CPM system.
70 OPH (18)	18 06 +02 30	DS	4.2, 6.0 K0V, K6	4"	! Famous yellow & red binary with 88-year period. "Justly famed as one of the most interesting of all binaries for amateur telescopes because of its rapid motion, large orbit & brightness." Minimum separation of 1.5" last occurred in mid-1989 with the ruddy companion SW of the yellow primary; the pair is now opening to another maximum of 6.8" in 2024 when it will be SE of it. Actual distance apart ranges from 12 to 35 x the Earth-Sun distance. "Celebrated DS." "Close, challenging binary." "Celestial dynamic duo." "Beautiful." "A superb pair." "Yellowish & rose-colored." "Yellow & purple." "Pale topaz, violet." "Bright yellow & orange." "Golden & rusty-orange, with a persistent hint of violet in the fainter star." A "Rapid revolver."! "Close, neat pair! Does swarm of 12th- thru 14th-mag. stars within 2' radius make an OC?" Truly "Beautiful color contrast of yellow & orange or violet." "Pretty orange-dwarf binary." "Very remarkable." "Presents puzzles & problems still unsolved" including suspected unseen third body in the system which may be planetary in mass! This lovely pair is

currently snug in a 3-inch - needs 6-inch & steady seeing to really enjoy. Just split in a 5-inch at 50X & neatly so at 75X, with tints quite obvious. Closeby: D=17LY!

LFT 1385 OPH (3)	17 58 +04 34	SS	9.5 M4V	---
M10 OPH (17)	16 57 -04 06	GC	6.6	15'
M12 OPH (16)	16 47 -01 57	GC	6.8	15'
M14 OPH (10)	17 38 -03 15	GC	7.6	12'

* Barnard's (Runaway) Star Red dwarf & closest neighbor to the Sun after Alpha Centauri system: D=6 LY! Also highest proper-motion/fastest-moving star in the sky: 10" of arc per year (or the apparent diameter of the Moon in 180 years!) due N. Also famous for possible presence of planetary system, which is on-again, off-again! Needs deep map (showing stars to 10th-mag.) & careful star-hopping to locate without use of setting circles. A "Remarkable star." "A mover!" Visible in 4-inch glass at 45X as a ruddy-orange speck of light suspended against the rich starfields of the Summer MW.

! Along with its near-twin M12 only 3 degrees away, the best of the great swarm of globulars in OPH. "Two of the finest anywhere in the sky - superb targets." "Great binocular pair!" "Wonderfully studded...one of the prettier sights aloft!" Very "Rich ...easily resolvable." "Beautiful cluster of extremely compressed stars." "This noble phenomenon is of a lucid white tint...clustering to a blaze in the centre." "Highly resolvable with 8-inch." "Central region pear-shaped with grainy texture" in 4-inch. "Silvery massed...suns." A "tendency to curved branches" of stars. "Two spectacular GCs." "An interesting pair, presenting a fine study in structural contrast." M10's "Tighter structure makes it more difficult to resolve (than M12)...the nucleus remains an unresolvable blaze" in an 8-inch. Both clusters equally distant (D=18,000LY) & lie about 4,000LY apart in space. Each of these vast stellar beehives would "appear as a bright naked-eye object of about 2nd-mag. as seen by the hypothetical inhabitants of the other cluster."! A marvelous sight in 12- to 14-inch apertures at 150X or more!

! Another beautiful stellar swarm! "Amateur scopes have a fairly easy time resolving separate suns...owing to the cluster's low star density." "Slightly larger but somewhat dimmer than M10, & has a looser structure...relatively slight central condensation." "A fine rich GC with a cortege of bright stars & many minute straggling outliers...a resolvable mass...condensed towards the centre with several very bright spots." Globe "Seems entangled in a long skein of stars lying roughly E-W." A "Stringy, bright, well-resolved halo at high X." "Slightly spiral; finely grouped." "Beautiful cluster." "Loose...easy to resolve." "Intermediate between the globular & diffuse galactic type. Distinguished surroundings." "Rich, fine, bright...& stars resolvable." M12 & M10 both look granular in a 4-inch at 80X & are centrally resolved in a 10-inch at 100X.

A fainter but much richer cluster than the previous two. "Most beautiful & delicate... of the finest star dust...excessively rich." "Lacks a sharp central condensation, the distribution of light being very smooth across the disc, with a gradual tapering off at the outer edges." Like the others, easily visible in 3-inch - but here "Large instruments are required to show the countless members appearing as if the whole field had been powdered with luminous dust." "A large globular of minute stars." "Large; glimpses of resolution." "Mottled look, partly resolves in 8-inch." "Of a lucid-white colour & very nebulous in aspect, which may be partly owning to its being situated in a splendid field of stars, the lustre of which interferes with it." "I can see the stars...this cluster is considerably behind the scattered stars, as some of them are projected upon it." "Quite impressive." "Conspicuously oval." "Extended - like blown star-dust." "Often passed over - what a shame."! A truly spectacular object in a 13-inch at 270X. D=33,000LY

M19 OPH (14)	17 03 -26 16	GC	7.2	14'

! "A mass of stars; large, fairly bright, but very low. Near the large blank space in the body of Scorpio, where Herschel found scarcely any stars" (his "Hole in the Sky"!). "A bewilderment of...clusters, accentuated by encompassing black rayless space-deeps." "Fine insulated GC of small & very compressed stars of creamy white tinge & slightly lustrous in the centre." "One of the more oblate globulars...its flattening of outline is noticeable even in very small telescopes." "Markedly oblate GC." "Beautiful cluster ...like a miniature Omega Centauri in 4-inch, central region grainy - on threshold of resolution." "Superb." But "Not especially well-placed for viewing." D= 30,000LY

M62 OPH (10)	17 01 -30 07	GC	6.6	14'

! With M10, brightest GC in OPH. "Located squarely on OPH-SCO border" & often listed under the latter. "Fine cluster, but very low." A "Miniature of M3 (in CVN)." "Striking cluster, dark nebulae abounding." "A fine, large, resolvable nebula; an aggregated mass of small stars running up to a blaze in the centre." "Unsymmetrical; in a rich field." "Asymmetric, with flattened SE region." "A distinctive northward 'bulge'...apparent." "Nearly a twin of M19 (4 degrees N). An impressive object." "Unusual irregular outline. The thickest massing of stars...is a 'perfect blaze, but not quite in the centre'...central condensation is actually 'well to SE of centre'...most 'comet-like' of all the M objects...color 'slightly bluish'." "Charming object." Very bright fuzz-ball in 4-inch glass - needs at least 8-inch for even partial resolution due to both the relative faintness of the individual cluster members & its low altitude above horizon. "Magnificent GC...set in an absolutely stunning star field." D=20,000LY

M9 OPH (10)	17 19 -18 31	GC	7.9	9'

Smallest of the Messier globulars in OPH & a most unusual-looking object. "Scraggly & uneven for a globular...looks like a very rich OC." "Small but scintillant & clear." "Small, apparently resolvable." "A myriad of minute stars, clustering to a blaze in the centre, & wonderfully aggregated with numerous outliers seen by glimpses." Like a "Miniature of M53 (COM)." "Impressive at 120X" in a 4-inch, but this object is really a big-scope globular needing at least a 12-inch for a decent view. D=26,000LY Two small faint globulars, NGC 6342 (3' & 10th-mag.) & NGC 6356 (2' & 8.4-mag.), lie slightly over a degree SE & NE, respectively. All three seen together in 6-inch RFT!

NGC 6572 OPH (12)	18 12 +06 51	PN	9.0/13.6	15"x12"

! Minute but intense, blue planetary like NGC 6210 in HER. "Tiny blue-green ellipse." "Tiny bright blue oval." "Very high surface brightness, bluish." "Small, very bright. Considered by Struve one of the most curious objects in the heavens." (It carries his designation Struve 6N.) "Slightly hazy." "A little elliptical, 5" or 6", bluish-green."

Apparent size does look much smaller than given in data column - more like 8" of arc. "Exquisite mite of a PN." "Clean greenish disc at 300X" on 8-inch. "Curious bright gaseous unit...bluish." "Extremely bright...perhaps the brightest of its kind." "The brightest of the nearly 50 planetaries (!) within the boundaries of OPH." "A beautiful little gem." The dim nucleus is all-but-invisible - at least partly "due to high surface brightness of engulfing nebulosity - an interesting twist to the phrase 'light pollution'." Fascinating sight in 6-inch at 100X - eerie-looking in 12-inch at 200X! D=1,900LY

NGC 6633 OPH (8)	18 28 +06 34	OC	4.6	27'		

Big, bright, scattered clan. Despite OPH's huge size, its only decent telescopic OC! "Very fine, with star 7th-mag. in large field; naked-eye object." "Superlative! Grand star-clouds following." "Sparse wide field cluster." "An elongated mass." "A lovely, great, straggling thing...of an absurd shape."! Some 30 to 60 members. D=1,000LY

NGC 6369 OPH (6)	17 29 -23 46	PN	11.5/14.7	30"	

Little Ghost Nebula "Dim ring-shaped, fainter cousin of Lyra's famed Ring Nebula (M57)." "Fine sight in small scopes." "Rather faint annular nebula." "Perfect ring." "Easy to scare up...a miniature Ring Nebula...an exciting PN...pale blue hue" evident in 4-inch on dark night. "Beautiful ring-shaped planetary." "A perfect smoke ring." "Green." Best seen in 8-inch or larger under steady, transparent skies. D=3,800LY

IC 4665 OPH (5)	17 46 +05 43	OC	4.2	41'	

A "Big dim glow." "Summer Beehive - superb." "Great binocular object." Lovely in a 4-inch RFT at 16X! Membership estimates run as high as 80 suns & total size as big as 70' of arc. A "Scanty sprinkling." "This big, beautiful object deserves to be in the Messier catalog." "Fine collection of about 30 stars...large, striking OC." D=1,300LY

Alpha ORI (6)	05 55 +07 24	SS	0.4-1.3 M1I-M2I	---	

! Betelgeuse Marking the "Giant's Shoulder" - the lovely radiant lucida of dazzling Orion, richest star-grouping in the heavens. "The California of the Sky!" & "Imperial constellation of the firmament." "Delights the eyes of all beholders." The star is "A most beautiful & brilliant gem! Singularly beautiful in colour, a rich topaz; in hue & brilliancy different from any other star I have seen." "Like a shiny copper penny!" If you think stars don't have color, just "Look at Alpha & Beta (Rigel) alternately to appreciate the contrast"! "One of the most striking stars...in the night sky - distinctly orange or red!" "Rare orange sapphire." "The Fire-Star." Truly a marvelous sight, whether using the unaided eye, binoculars or a telescope! A huge semiregular super-giant sun, whose size varies from 480 million to 800 million miles in diameter as it pulsates over an approximate period of 5.7 years! Has two very close companions, the inner one completing an orbit in just two years & actually lying *inside* the giant star's outer atmosphere (in which water vapor - steam - has been detected)! Displays a very "brilliant spectrum" for those observers using eyepiece spectroscopes. D=520LY

Beta ORI (16)	05 14 -08 12	DS	0.1, 6.8 B8I, B5V	10"	

! Rigel "Leg of the Giant." A beautiful blue-white supergiant sun with a close-by, fainter companion forming a splendid magnitude-contrast pair visible in a 3-inch glass at 90X. The secondary sits in the primary's glare & needs steady seeing (typically infrequent in Winter, common in Fall) for a clean split in apertures under 6-inches. A superb sight in 12- to 14-inch instruments! "I always see a blue tinge in the great star, resembling that of Vega. Beautiful object, & fair test for small telescope." The little star has been described as pale red, orange, azure, sapphire blue, bluish & just white in hue. "A boy blue with a companion on his shoulder." "Brilliant." "Splendid object." "A true supergiant, a blazing white-hot star of intense brilliance & dazzling beauty." Fixed pair showing little change, but surely a physical system! The companion itself has been reported to be an exceedingly close pair (7.6-mag. & maximum separation of just 0.1"), visible only in largest refractors (Lick 36-inch, Yerkes 40-inch). D=900LY

Eta ORI (8)	05 25 -02 24	DS	3.1-3.4, 4.8 B1V+B2, B?	1.5"	

! Very bright, tight binary - test for 4-inch & not easy in 6-inch unless seeing is good. "White, purplish. Excellent test." A "Brilliant pair." "White/blue; 500X splits it nicely" in an 8-inch. So does 290X with a 13-inch. Resolvable at around 150X in a good 4- or 5-inch glass. "Helium stars involved in calcium clouds." A "Complex multiple-variable." This pair forms a radiant big-scope binary having a very long but undetermined period; the primary is also an eclipsing binary with 8-day orbital period. There's an unrelated 9th-mag. field star at a distance of 115" to the SE. D=1,400LY

Lambda ORI (15)	05 35 +09 56	DS	3.6, 5.5 O8III, BOV	4"	

! Neat off-white-hued duo in Orion's head. "An elegant DS for the small telescope." "Many & curious are the reported discrepancies in the colors of this pair, all the more strange since both stars are type O & should appear simply white...the pair has always appeared (to me) sparkling white with just a hint of a light amber tint." A "Lovely contrasting pair." "Yellow & purple." "Pale white & violet." "Striking white & orangish." "Yellow & red." "Blue/blue. 20+ members over 1 degree wide area form OC COL 69" - known as the **Lambda Orionis Cluster**. The double is relatively fixed & likely forms an ultra-long-period binary system. Lovely sight in a 5-inch at 75X & especially pretty in larger apertures. Quite remote for so bright a double: D=2,200LY!

Delta ORI (10)	05 32 -00 18	DS	1.9-2.1, 6.3 O9II, B2V	53"	

! **Mintaka** Wide, easy magnitude-contrast pair for low powers. "Seems tailor-made for binoculars" & RFTs! "Greenish white, white; (& again) pale green, pale violet or lilac." "Coarsely double. Brilliant white, pale violet." "Yellowish white & reddish. Bright, beautiful." "Bluish companion." The "greenish; slightly variable" primary is another eclipsing binary with 5.7-day period. Sweet in a 3-inch at 30X, but looses its appeal as magnification increases (& field of view decreases!). Delta sits right on the Celestial Equator at zero degrees DEC - as result, often used to align traditional setting circles & also planetarium optical projectors. A fixed physical system. D=1,400LY

Zeta ORI/NGC 2024 (16/7)	05 41 -01 57	DS	1.9, 4.0 O9I, BOIII	2.5"	
		DN	---	30'	

! Radiant physical pair with blue-white tints. Often a tough split due to closeness of companion & typical poor Winter seeing. They can be resolved with apertures under 5-inches on a good night, but were "Singularly missed by (William) Herschel...seems of some nondescript hue, about which observers do not agree. Struve calls it 'olivacea-subrubicunda' (slightly reddish-olive)!" Colors also given as "topaz yellow & light purple", "yellow & blue", "both bluish" & both "brilliant white." "So remarkable & elegant a pair." "Pretty." A distant 10th-mag. star at 58" forms rather weak but "Fine triple." A-B's period "must be at least several thousand years." A big scope double - stunning in a 12-inch on a steady night! D=1,400LY Zeta "Lies in a remarkable field of...nebulosity which is excited to luminosity by the star." Just 15' E & slightly N is NGC 2024, the big **Flame Nebula**, a "Glowing mass of bright & dark areas...split by a wide dark lane running from N to S." "A complex, mottled gray patch of light" the size of the Moon. Its tree-like shape can be glimpsed in a 4-inch glass at around 90X (especially if the glare from Zeta is reduced by placing the star itself just outside of the field of view). This "pale nebulosity" is also called the **Burning Bush Nebula**. A very "faint reef of nebulosity" extends S of Zeta to the Horsehead Nebula (see below).

23 ORI (4)	05 23 +03 33	DS	5.0, 7.1 O9II, B2V	32"

An overlooked, wide easy combo for small glasses. "Greenish white, white; (& again) pale yellow & fine blue. Beautiful color." "White, pale grey. Neat DS." A relaxing change of pace after tackling close pairs like Eta & Zeta! Nice in 2.4-inch at 25X.

Sigma ORI/Struve 761 (16/2)	05 39 -02 36	DS	4.0, 10.3, 7.5, 6.5, O9V, , B2V, B2V	11", 13", 43"
		DS	8.0, 8.5, 9.0	68", 8"

! An amazing colorful multiple-star system - a quadruple & a triple! And all visible in a 4-inch glass! Sigma "Remarkable...containing five visible components." (A is an ultra-close 0.2", 170-year period binary, making this actually a quintuple system!) "A beautiful little triangle...all white" describes Struve 761, which lies 210" to the W of Sigma. "Beautiful multiple star, many colors contrasted." Some descriptions cite Sigma as a triple also, depending on how many components the observer saw. "A sextuple star, which appears as a double triple." "A double-treble star." This clan has also been dubbed a "Double-quadruple" & as "Showing as many as 10 stars in large telescopes."! But others completely overlook the Struve trio, seemingly unaware of its presence - only a "pretty quadruple" being mentioned. "In a fine group of two pairs." Sigma (itself) is called "Perhaps the most impressive & colorful of all Orion's stellar treasures. To complete the picture, in the same telescopic field of view is the faint triple Struve 761. Altogether an extraordinarily rich & unexpected sight, to be returned to again & again." "A very remarkable multiple star." "Fine group with striking colors." Appears ever more lovely as aperture increases & these contrasting tints become unmistakably obvious. White, yellowish, orangish, red, grape red, ash, dusky & pale grey are among the heavenly hues having been reported here by various observers! Incredibly, *all one vast physical system*! "Fantastic field." D=1,400LY

Iota ORI/Struve 747 (11/5)	05 35 -05 55	DS	2.8, 6.9 O5, B9	11"
		DS	4.8, 5.7 B1, B1	36"

! Two pretty diamond-like pairs in the same radiant gem-field, just 8' apart & forming what can be viewed as a very wide & diverse "double-double" - or four-sun quadruple system! Both doubles are easily resolved in a 3-inch glass at just 30X. Iota resembles a miniature (fainter) version of Rigel. "Gem-like dazzling primary in excellent field, white & blue." An 11th-mag. at 50" distance "makes this an easy & attractive triple star for the small telescope." (But "small" here means something in the 8-inch range due to this object's dimness.) "A fine triple...white, pale blue, grape red." Struve 747

....		(Please see Page 96)			
....					

is a wide, equal "fine white pair" whose tints have also been described as "yellowish & ashy." A faint third double (Struve 745, 9th-mag. suns set 29" apart) sits closeby. A subtle nebulous glow can often be glimpsed here on dark nights, especially around Iota. All three pairs lie in same wide field as the Orion Nebula (below)! D=2,000LY

W ORI	05 05 +01 11	SS	6.2-7.0	---	
(0)			N5 (C5)		

! Another of the reddest stars in *Sky Cat 2000.0* - one whose glow may help warm you up on a cold Winter's night at the telescope! "Very deep orange; you can almost feel the hot wind in your face!" A semiregular pulsating carbon star with 212-day period.

Theta-2 ORI	05 35 -05 25	DS	5.2, 6.6	52"
(2)			O9V, B7IV	

A roomy easy double in the same eyepiece field with Theta-1 ORI & the Orion Nebula complex. "A wide white pair in overwhelming field." "Coarsely double." An "Easily split" combo in binoculars. Part of the amazing "Trapezium Cluster" clan (see next)!

Theta-1 ORI/M42/M43	05 35 -05 23	DS	6.4, 7.9, 5.1, 6.7	9", 13", 22"
(15/21)			B0V, B0V, O6, B0V	
		DN	4.0/9.0	66'x60'/20'x10'

! **Trapezium/Orion Nebula** The finest diffuse nebula in the sky & perhaps the most magnificent deep-sky showpiece of them all! Beautiful embedded multiple system also ranks as one of best of its class - like diamonds on green velvet! Together an amazing spectacle in all size & types of telescopes. "One of the most wonderful objects in the heavens; readily visible to the naked eye. The telescope shows an irregular branching mass of greenish haze, in some directions moderately well defined where the dark sky penetrates it in deep openings: in others melting imperceptibly away...singular convolutions...a prodigious diffusion." "I do not know how to describe it better than by comparing it to a curdling liquid, or a surface strewed over with flocks of wool, or to the breaking up of a mackerel sky." "Most beloved of all emission nebulae." "Has inspired more adjectives than any other deep-sky object...aesthetic shock."! "Most magnificent of the nebulae." "A convoluted veil." A "Greenish irregular fan-shaped mass." "Most celebrated...gigantic nebula of gas & dust from which a star cluster is being born...indisputedly the finest diffuse nebula in the sky...wreaths & swirls of gas become more complex & more breathtaking the larger the aperture...appears distinctly greenish...dark lane known as the 'Fish Mouth' separates M42 from M43 - a smaller & rounder patch that's really part of the same cloud" as M42 - an "Irregular, comma-shaped nebulosity" just 7' to its N. "A veritable, multi-course, gourmet feast for the telescopist. Defies a brief description! Find it - revel in the wisps, filaments & colors - & come up with your own!" "A stupendous mass of gas...a gigantic whirlpool...a glorious & wonderful sight...words fail to describe its beauty...wonder of the skies." "Resembles a giant luminous fan." "A wondrous mass." "This wondrous beast." "An awe-inspiring sight." "Grandest...tremendous greenish mist...wondrous...a distinctive turquoise cast. Magnificent sight - finest photos can't convey same thrill as viewing M42 firsthand." "Undoubtedly the finest & most detailed of all the nebulae N or S." "An inconceivably great cloud of cosmic dust driven to & fro in never-ceasing currents." "Most primitive form of matter known...in seething & well-nigh chaotic turmoil." "An unformed fiery mist...the material of future suns." "One of the youngest nebulas in the sky...more than 500 irregular variables and faint flare stars. One of

....
....
....
....
....
....
....
....
....
....
....
....
....
....
....
....

(This grandest of deep-sky objects - aside from the Milky Way itself - warrants longest description!)

....
....
....
....
....
....
....
....
....
....
....
....
....
....
....
....
....

the most remarkable areas in the heavens...greenish glow fills entire field, (the) 'Fish's Mouth'...a very conspicuous dark wedge - from both sides...great luminous bands give way, forming a ring that can be traced through nearly its full circumference under favorable conditions. M43 is a roughly triangular detached bright patch crossed by a dusky streak." "Quickly explodes into a glowing cloud of great intricacy when viewed with any optical aid. It looks to me like a cupped hand with tenuous, glowing fingers extending from the main body of the nebula toward the myriad of field stars." "One of the most wondrously beautiful objects in the heavens...impressive beyond words... draws exclamations of delight & astonishment from all who see it. The great glowing irregular cloud, shining by the gleaming light of the diamond-like stars entangled in it, makes a marvelous spectacle...unequaled anywhere else in the sky. Barnard found it resembling a great ghostly bat...& spoke of a feeling of awe & surprise each time he saw it (in the great Yerkes 40-inch refractor). To many others it creates, as does no other vista of the heavens, the single overpowering impression of primeval chaos, and transports the imaginative observer back to the days of creation. This irresistible impression is more than a poetic fancy...for the Orion Nebula is undoubtedly one of the regions of the heavens where star formation is presently underway." "The supreme wonder of all supernal wonders. Delicate emerald. The famous Trapezium, comprising four involved giants...thrilling beyond words!" "Here is found the fascinating colored multiple star, Theta-1 Orionis, known as the Trapezium. The nebula itself is greenish & of irregular form - much detail with branches, rifts, & bays; the entire nebula with its stars repays long observation." "Probably the best known multiple star in the sky, & one of the most interesting for the small telescope...a favorite of all observers...the bright core of a compact cluster of faint stars." A 2.4-inch glass easily shows the 4 primary stars, while a 4-inch reveals two others (both of 11th-mag. & 4" distant from the first & third brightest Trapezium members). "A sextuple star in the heart of the Great Nebula" near the tip of the Fish Mouth. Color contrasts among the 6 stars become evident in a 6-inch. An "Incredible, beautiful multicolored foursome." "Pale white, faint lilac, garnet, & reddish", blue also being among the hues given later by the same observer. "White, lilac, garnet, reddish, &...." "But (the) impression of reddish tints in any of these (four brightest) stars must be attributed to the effects of contrast with the greenish background of the nebulosity." "White quadruple system." "These stars...shine with a fiery blue-white intensity against the soft, gossamer clouds of the nebula itself." Third & fourth brightest Trapezium stars are eclipsing binaries with periods of 65 & 6.5 days, respectively. "Frequently, observers fail to distinguish M43...as a separate object...while (it) was discovered telescopically as early as 1610, M43 was not recognized as a separate entity for another 121 years. Mere words cannot do justice to this magnificent pair. Even the finest drawings & photographs cannot capture the thrill of viewing M42-43 visually. What a view!" A 3-inch glass at 45X reveals darker areas & long filaments in this bright green cloud, while 6- to 8-inchers at 100X strikingly shows its "curdled" or mottled structure. Apertures in the 5-inch & above range will also show faintly glowing rosy-pink areas in the nebulosity itself.

Many faint red stars can be glimpsed sprinkled about the nebula in larger scopes. The author has actually seen the Orion Nebula *projected in color* onto a sheet of glossy white paper through a 30-inch refractor! To see full extent & glory of this "Greatest of all spectacles" place it just outside the eyepiece field & let it slowly drift into & across view - using averted vision to see faint details & direct vision to pick up colors. You'll then sense its "Overpowering sublimity" & "Incomparable...splendor."!! D=1,600LY

M78 ORI (10)	05 47 +00 03	DN	8.0	8'x6'

A "Singular 'wispy' nebula." "Famous cometlike form, with broad...fanlike tail extending to the S. Two nearly identical 10th-mag. stars pose as 'nuclei' to complete the comet illusion." "Weird gaseous nebula. N edge clean-cut, S portion diffuse. Interesting!" "Like breath on a mirror." "Faint concentration with definite 'combed' structure." "Oval haze." A "Featureless reflection nebula." "Shows very little detail to the visual observer." "Bluish. One of the sky's brightest reflection nebulae & easiest to observe." Visible in a 3-inch at 30X, it appears "Comet-shaped with apex directed northward" in an 8-inch at 80X. "Triangular, with...double & other beautiful features. Worth study" in 10-inch. The comet's 'tail' blows southward, where "it melts away" into the dark sky. "The sky surrounding M78 seems to have a misty sheen" caused by several other fainter nebulous patches lying within the same field, each having its own NGC number. "One of my favorites...has curious 'ink blob' symmetry." D=1,400LY

NGC 2022 ORI (4)	05 42 +09 05	PN	12.4/15	18"

Dim ring-shaped planetary near Orion's head for 6-inch & larger scopes. "Small, faint & distinct with annular form." "Clear bluish" in 10-inch. "Curious." "Of a bluish white tint. Small & pale, but very distinct." "Diffuse edges." "Rather oval & perhaps of a mottled light." "Small PN, in starry field." "Although Orion overflows with vast regions of bright & dark nebulosity, it can lay claim to only a single PN that is bright enough to be seen in most backyard telescopes...difficulty detecting any color...detail requires a medium- or large-aperture telescope & high magnification. Only then, and by using averted vision, will the cloud's ringlike structure be seen." D=7,000LY

NGC1973-75-77/	05 35 -04 35	DN	----	40'x25'
(4/2) NGC 1981 ORI		OC	4.6	25'

Large, faintly glowing nebulous complex & bright scattered OC of a dozen or so stars in the same wide field. "Elongated nebulosity just N of the Orion Nebula centered on 5th-mag. star 42 ORI. Object would be more celebrated if it were not so overshadowed by M42." "Often neglected." "Extensive...with several bright stars." "A large, bright, bluish arc of nebulosity stretching between 42 ORI & two dimmer stars." "Clearly seen through binoculars. Warm, irregular form engulfing 4 stars." Just N is the OC itself, strangely included by Pickering as one of his 60 finest objects even though it's anything but spectacular!? "Brilliant field...grand neighbourhood; sweep well over the whole space." Includes the close DSs Struve 743 (mags. 7 & 8, sep. 2") & Struve 750 (mags. 6 & 8, sep. 4"). This wonderland needs a WF & dark night for a good view.

COL 70 ORI (1)	05 36 -01 00	OC	0.4!	150'

* Epsilon Orionis (Belt Stars) Cluster Stunning example of the many wide, loose aggregations of stars scattered across the heavens - in this case over a hundred of them centered on Epsilon in a circular pattern. Best in binoculars & RFTs - an amazing sight in a 4-inch RFT at 16X! Epsilon's name - Alnilam - means "A Belt of Pearls" referring to the three belt stars, as does the designation COL 70. "An OC that all of us have seen but few are aware that it is a cluster...surrounds & includes all three stars in Orion's belt. In all, 100 suns spanning 3 degrees belong to this wide group." (The magnitude listed is that of the integrated light of all these stellar jewels *combined*!)

Barnard 33/IC 434 ORI	05 41 -02 28	DN	---	6'x4'	
(6)		DN	---	60'x10'	

* Horsehead Nebula Best-known & most spectacular dark nebula in the sky as seen in photos - but sadly also one of the most difficult objects to glimpse visually in the telescope! Extending southward from Zeta ORI for 1 degree "Is a strip of nebulosity into which is indented (about halfway along its length) the celebrated Horsehead Nebula...notoriously difficult to see with amateur telescopes." "A dark nebula (B33) superimposed on a very faint emission nebula (IC 434)...use H-Beta filter." "Requires dark sky, nebula filter." "Everyone's favorite dark nebula." "IC 434 could be seen as a faint 'foggy' wall with a perfectly straight E edge; the surprisingly large Horsehead requires absolute concentration to be glimpsed" - from an observation made using a 12-inch & nebular filter. Experienced observers have glimpsed it in apertures as small as 5- & 6-inches without a filter, & there's at least one report of it being picked up in 10X70 binoculars! But more typically: "Nebula filter - clear steady skies - Orion near culmination (on the meridian) - minimum 8-inch aperture - wide field (greater than 1 degree) - 50X to 100X - dark adaptation - sense of scale - ALL necessary to see this 5' beast."! So in no case is this object to be considered a *visual* showpiece! Yet: "One of the most extraordinary objects in the sky." It was exuberant descriptions like this of objects such as the Horsehead causing early disappointments & frustration at the eyepieces of small telescopes (while other truly wonderful sights - often nearby - went ignored) that led the author more than four decades ago to undertake the visual survey of the sky that ultimately led to this compilation of heavenly showpieces. The noted deep-sky observer Walter Scott Houston warned me way back then that this was truly "an impossible project in aesthetics" but still wished me every success in my efforts!

Epsilon PEG	21 44 +09 52	DS	2.4, 8.5	143"	
(4)			K2I		

Enif/Pendulum Star Another of those rare deep-sky objects like Polaris (UMI) & the Blinking Planetary (CYG) that 'does something' in the eyepiece while looking at it! "This object, when near the meridian, will exhibit a phenomenon, noticed by John Herschel - the pendulum-like oscillation of a small star in the same vertical with a larger one, when the telescope is swung from side to side. This, he thinks, is due to the longer time required for a fainter light to affect the retina, so that the reversal of motion is first perceived in the brighter object. I have seen this strikingly in Delta & Zeta ORI, & Delta HER." This effect can also be seen by gently tapping the telescope tube (especially on a shaky mounting!) & is easily visible even in a 3-inch glass at just 45X. "Contrasty colors." "Yellow & violet." "Yellow supergiant with...wide bluish companion." The brighter star is "Suspected of variation" & a flare-up to magnitude 0.7 was actually reported in 1972! Wide optical duo - primary's D=780LY

M15 PEG (20)	21 30 +12 10	GC	6.4	12'	! "Rich, compact globular." "Outstanding bright GC...nearby 6th-mag. (star) serves as sure guide to its location...(a) glorious misty sight in an attractive field. With apertures of 6-inches or so its outer regions can be resolved into a mottled ground of sparkling stars & larger apertures show stars all the way to the bright condensed core." "Beautiful globular with intense core & well-resolved halo." "Bright & resolvable, blazing in centre; a glorious object." "A 'noble cluster'; a fine, large, bright globe of stars." "A pile of diamonds on black velvet! 14th-mag. 1" (diameter) PN - I'll be hanged if I can find it!" with an 8-inch. Known as Pease-1, it's been glimpsed in big scopes using the technique of "blinking" with a nebula filter. "Only known GC with a PN...grand globular...nestled in a fine star field, center very intense...faint extensions." "Striking condensed GC...3-D effect in 12-inch & larger telescopes." "Magnificent." "One of the richer & more compact globulars, remarkable for the intense brilliance of its central core where the countless suns seemingly crowd together into a blazing nuclear mass." Beautiful in 6-inch, but still not completely resolved even in 13-inch most nights. "Astounding view in 17-inch...breathtaking in 36-inch." D=34,000LY
NGC 7331 PEG (5)	22 37 +34 25	SG	9.5	11'x4'	"An alluring SG prominently visible in backyard scopes." A "Large, bright spiral." "An elongated smudge in apertures of 4-inches or so." "Almost edge-on. Dust lanes visible." "Strikes most amateurs as a tiny version of the Andromeda Galaxy." Nice in 8-inch at 80X. "Very large, elongated spiral with very faint companions." A "Dimly glowing N-S streak." In same field, 30' to SSW is the "Famous **Stephan's Quintet** clan" - a "Tight little group of remote galaxies." Needs 10-inch & dark skies (has been glimpsed in 3.5-inch!). Mags. range from 13th to 14th. NGC 7331 D=50,000,000LY
Alpha PER/MEL 20 (4)	03 22 +49 00	AS	1.8/1.2 F5I	185'	* **Mirfak/Alpha Persei Association** Here's the finest example of a stellar association anywhere in the N sky! Much too spread out for normal telescopic fields, this group is a spectacular sight in binoculars & RFTs which "Reveal a brilliant scattering of stars." "Splendid treat for binoculars surrounding Alpha PER! 100+ members, 5 degrees extent." "Princely Mirfak throned in the midst of a truly royal council of stars" flowing SE of it along the course of the MW. "Many are set in close-knit duets, trios & quartets, & add greatly to the group's visual impact. Brightest of all is...Mirfak, the most brilliant gem in this celestial diamond mine. Literally explodes with beauty. Many stunning blue-white starry jewels." "A fine field of stars which form a rich & brilliant group for small scopes...truly splendid." Mirfak a "brilliant lilac." D=560LY
Eta PER (13)	02 51 +55 54	DS	3.8, 8.5 K3I, B?	28"	! A neat color & magnitude contrast pair with other faint stars close-by. An "Orange supergiant with blue companion forms an attractive DS for small scopes...field of view contains a sprinkling of background stars." "Sparkling background." Has "Five faint comites." "Beautiful DS, blue & gold, resembles Albireo." "Orange, smalt blue; the colours in clear contrast." "Very yellow, very blue." At 67" is a 10th-mag. companion star which careful inspection will show to be a close, equal pair 5" apart. AB lovely in a 5-inch glass at 50X - other attendants need at least a 6-inch at 90X to be well seen.

Beta PER (8)	03 08 +40 57	SS	2.1-3.4 B8V+G5IV	---	*** Algol/Demon Star** Prototype & best-known of the eclipsing binaries. Every 2.9 days one of the stars in this visually-unresolved pair passes in front of the other, causing a dimming (or 'winking' as the ancients termed it) obvious to the unaided eye. To see it, just compare Algol's changing brightness with the constant light of nearby Mirfak. The entire eclipse from start to finish lasts just 10 hours, 2 of them at minimum light. "One of the most celebrated variable stars in the sky." While not a telescopic show-piece as such because of the time-span involved in seeing its magnitude fluctuate, this object is still fascinating to monitor whenever visible. A "Remarkable star." Varies as "Regularly as clockwork." Here we're watching two giant suns orbiting each other far out in the depths of space - with no instrument other than the naked eye! D=100LY
Epsilon PER (11)	03 58 +40 01	DS	2.9, 7.6 B0V, A2V	9"	An interesting but somewhat challenging pair for small scopes. "Not an easy double." "Companion difficult to see...because of magnitude contrast." It definitely does seem fainter than its listed value. "Green, blue-white. B supposedly changes color, blue to red." Changes in brightness too? "Pale yellow & slate-coloured." "White & blue."
M34 PER (17)	02 42 +42 47	OC	5.2	35'	! A "Very grand low-X field, one of the finest objects of its class." "Quite distinct & pretty, looking like a 'box of jewels'." "Overshadowed by the more spectacular Double Cluster (see below)." "A celestial aegis hung aloft in splendor!" "Outstanding OC...a grand target for binoculars...wonderful...to view through telescopes & binoculars alike ...explodes into stardust with only the slightest optical aid...the group as a whole looks rectangular...some observers comment about strings of stars that run across the object." "A scattered but elegant group." "Beautiful." A lovely sight in small glasses, but it is "Not more spectacular in large telescopes, as it does not seem to have the needed fainter stars to buttress the view." "Far less rich & condensed than the Double Cluster ...about 80 stars splashed over an area approximately the size of the full Moon." "Very bright & large, visible with the naked-eye" on dark nights. "Loose cluster...very fine... main stars form a distorted 'X'...includes Otto Struve 44" - a tight (1.4") 8th-mag. DS located SE of the center of the cluster. There's also a much wider (20" apart) & easier 8th-mag. pair near the center itself, designated Herschel 1123. D=1,500LY
M76 PER (14)	01 42 +51 34	PN	11.5?/17.0	140"x70"	! **Little Dumbbell/Barbell/Cork Nebula** Although generally considered to be the faintest object in the Messier catalog, this "Diminutive planetary" is visible in a 3-inch glass & is a very fascinating sight in a 6-inch. "Pearly white nebula, double, curious miniature of M27." "Described by many observers as rectangular, oriented approxi-mately N-S...(to one) it looks like a celestial peanut. Larger instruments display M76 as two oval spheres of grayish light that appear to touch one another." "Unusual two-lobe structure." (It's this double aspect that led to M76 having two NGC numbers: NGC 650 & NGC 651.) "A rewarding object - a miniature of the Dumbbell (looks more like one than M27 in small telescope)." "Somewhat resembling a cork." Also called by some the "Butterfly Nebula"! "Radial filaments at high X" seen in 8-inch. "Double spiral nebula, joined, with ring." "Small nebula with two distinct nuclei."

"M76 12th-magnitude? I find this hard to believe, & feel it's at least a magnitude brighter." And so it seems - one observer even rates it as 10.1! "Large PN." "Indeed a treasure worth the hunt." "Exciting (in 8-inch)." "Marvelous (in 12-inch)." "In an intensely rich vicinity" - causing it to seemingly 'float' before your eyes! D=4,000LY

NGC 869/	02 19 +57 09	OC	3.5	30'	
NGC 884 PER		OC	3.6	30'	
(21)					

! Double Cluster Amazing dual stellar jewelboxes easily visible to the naked eye as a misty elongated glow in the Perseus MW - & as a glittering starburst in binoculars! Together they rank as the finest OC for small telescopes & are truly a superb sight even in large instruments (in which typically only one at a time can be fitted into the field of view). NGC 869 (the W member) is the richer & more impressive of the two, displaying many contrasting colors & a striking "Star Tunnel" effect upon looking into its sparkling core (mainly evident in 6-inch & bigger scopes)! "NGC 869 brighter & richer of pair (200/150 stars - some counts give 350/300 stars)...small scopes have the advantage...for with low X both clusters can be seen in the same field of view. Most of the stars are blue-white but there are several red stars to be spotted among the masses of glittering points...breathtaking sight in all apertures." "Two gorgeous clusters...visible to the naked eye as a protuberant part of the Galaxy. With 64X these superb masses were visible together...a ruby & a garnet in 884. The red stars are all associated with 884." "Affording together one of the most brilliant telescopic objects in the heavens." A "Grand sight in binoculars...a little too big for an 'all at once' view in a telescope." "Two magnificent clusters." "Famous...a wonderful object...one of the finest clusters for a small telescope. The field is simply sown with scintillating stars, & the contrasting colors are very beautiful." "Spectacular stellar aggregation...a delight to beginner & veteran observer alike...magnificent assemblage...simply behold its beauty...gazing at these clusters produces a succession of feelings too subtle & too complex to be captured by words...a unique deep-sky object." "Stunning!" "One of the most marvelously beautiful objects in the sky...dazzle the eye with their lively beams." "The preceding cluster (869) is rather the finer of the two...neither taken separately is the equal of M41 (CMA), nor M35 (GEM)."!? "Either one is a splendid group & under good observing conditions, the pair is one of the most brilliant & spectacular objects of its kind." NGC 869 "A glorious cluster...brilliant mass of stars... fills the whole field of view & emits a peculiarly splendid light. In the center is a coronet, or rather ellipse, of small stars" & NGC 884 "Another gorgeous group of stars ...(a) sprinkle." "Much prettier looking in a smaller telescope than a larger one. There seems to be a 'hole' in the center (of 884) where few stars are visible. Distinctly reddish variable stars in both clusters - the conspicuous red stars are all found around NGC 884." When "Taken together they provide us with a view that overflows with splendor...even the smallest of instruments begins to unleash the beauty that is the Double Cluster...a fiery radiance against the star-filled backdrop...stars shining with subtle hues of yellow & red could be seen scattered throughout a field dominated by blue-white suns." "One of the truly classic examples of a galactic (open) cluster, & a wonderfully beautiful object. Among the all-time favorites for amateur observers...the pair forming one of the most impressive & spectacular objects in the entire heavens. The bright members, which present such a glittering spectacle in the small telescope, are all great blazing supergiants of almost unimaginable brilliance." Their centers are about 1/2 degree apart on the sky & there has been much uncertainty about whether they are gravitationally linked. "Two splendid naked-eye clusters...overlapping so as to appear physically connected." But "Not physically related!" "A great galactic coincidence." The latest word seems to be that they "Do not lie right next to each other in real space, though they are close" (D=7,200LY & 7,500LY); & so these starry flocks are likely in slow orbit about one another as they fly across the heavens together. Many contrasting star colors are evident here even in a small glass, including a fine ruby star between the two groups. Strangely, "Messier apparently missed it" - & many question why, since it's obvious even to the unaided eye. (The argument that it's *too* obvious to be mistaken for a comet - Messier's purpose in compiling his list - doesn't hold up, since he *did* include the even more prominent Beehive & Pleiades clusters!?) "Tremendous blaze of scintillating suns...commanding." Superb, 6-inch RFT at 24X!

NGC 1245 PER (3)	03 15 +47 15	OC	8.4	10'	A "Gorgeous OC." "A very faint large cloud of minute stars...beautifully bordered by a brighter foreshortened pentagon." "In midst of the Alpha Persei Association - brilliant but tiny." "Congeries of low-mag. stars." "150-plus 12th-mag. members; 2 brighter stars in foreground; very rich." "It has a gathering spot, about 4' in diameter, where the star-dust glows among the minute points of light...elegant sprinkle." "The large stars are arranged in lines like interwoven letters." Nice sight in 8-inch at 60X.
NGC 1528 PER (8)	04 15 +51 14	OC	6.4	24'	"Bright cluster, good low-X object." "Large cluster, use low X - 75 stars in 6-inch." "Gorgeous...glows like a gem" in 5-inch RFT at 20X. "Radiant cluster - entire zone truly inspiring." "Large but sparse for its size." Like NGC 1245 above, this clan is usually overlooked because of the Double Cluster (& also M34) stealing the show!
Alpha PSC (19)	02 02 +02 46	DS	4.2, 5.1 A0, A3	2"	**! Alrescha** Tight pair with weird subtle tints! "Greenish white, blue. I found contrast certain, but (primary) troublesome as to color, usually ruddy, or tawny, sometimes blue. Pale yellow, brown yellow, 'quite satisfactory' (in) 3.7-inch...pale yellow, tawny or fawn color, 'certain' (in) 5.5-inch...'No strong contrast'...brownish, at first fancied bluish, (in) 9-inch."! "Challenging DS...colours bluish-white although some observers see the brighter star as greenish." "Pale green, blue; splendid object." "Green-white; blue or brown. Difficult in 3-inch; color contrast remarkable." "Weird coloring." "Curious tints, somewhat greenish." "Blue-white & yellow." Although both stars look white in a 13-inch at 145X, smaller apertures definitely give an impression of color. A binary with 720-year period, the two suns are slowly closing to a minimum of 1.0" in 2074 (last widest around 1700 at 4"). Both stars are spectroscopic binaries. D=130LY
Psi-1 PSC (8)	01 06 +21 28	DS	5.6, 5.8 A1V, A0V	30"	Easy matched duo for small glasses, both blue-white. "Neat pair - silver-white stars." "Yellow & pale blue." "Nice separation, equal brightness." "Splendid low-X field."

Zeta PSC (10)	01 14 +07 35	DS	5.6, 6.5 A7IV, F7V	23"	Another wide pair - this one a pale yellow & pale lilac combo. "Easily split bright couplet." "Visible in smallest telescopes." "Wide...blue-white." "Colorful pair." "Yellow, pale blue." "Yellowish, pale lilac or rose." "Both pale yellow." "Silver white, pale grey - fine & easy object." "Yellowish & orange, color contrast & separation both very nice; both may be variables." "White & yellow-white; not terribly exciting!" in 8-inch. Striking in 3-inch at 30X. "Beautiful." CPM pair. D=140LY
55 PSC (11)	00 40 +21 26	DS	5.4, 8.7 F3V	6"	A magnitude & color contrast pair for medium apertures. "Very yellow, very blue... (companion) very small, color indistinct" as seen in 3.7-inch. "A beauty." "Striking." "Nice colors." "Orange & sapphire blue." Best viewed in 6-inch or larger instrument; challenging for a 3-inch glass. But in an 8-inch, a "Neat & easy pair!" CPM system.
65 PSC (9)	00 50 +27 43	DS	6.3, 6.3 F4III, F5III	4"	Perfectly equal pale-yellow duo. Just split in 2-inch glass at 50X - neat in a 5-inch at same X. "Evenly matched pair." "Yellow & blue-white." "Close." CPM double.
19=TX PSC (5)	23 46 +03 29	SS	4.5-5.3 N0 (C7)	---	! Reddish-orange semiregular variable with 220-day period, located in the 'Circlet of Pisces' asterism (itself "an attractive sight" with the unaided eye). Has a very red color apparent in all apertures, including binoculars. Its "color is excellent." "The brightest N-type star in the N Hemisphere." "Magnificent object." "Decidedly deep orange; quite alone in a dark field." "To appreciate the unusual ruddy tint of the star... compare its light with the nearby bluish stars 21 & 25 PSC" to its SE. D=400+LY
Alpha PSA (2)	22 58 -29 37	SS	1.16 A3V	---	**Fomalhaut** The "Fish's Mouth." "Appears as a glittering object, low down in the S, below Aquarius." Also known as "The Solitary One", the "Supremely Lonely Star" & "The Loneliest Star" due to its isolation in a very blank area of sky. "A very sightly Autumn visitant from the southland, the 18th in order of celestial brilliants." Appears a lovely radiant, blue-white tint both to the unaided eye & in the telescope. Strangely, at least two classic observers of the past refer to it as "reddish" in color!? D=25LY
k/Kappa PUP (10)	07 39 -26 48	DS	4.5, 4.7 B6V, B5IV	10"	! A superb, bright easy double resembling Gamma ARI & sometimes referred to as its southern-sky twin! "Striking...near-identical blue-white components easily divisible." "Beautiful white couplet." Easy in a 2.4-inch at 30X & stunning in all size scopes - like a pair of 'eyes' staring at you! "Both blue-white; cosmic headlights!" "White or pale yellow." "Both topaz yellow, but the tinge...may be owing to its low altitude" as seen from the British Isles. "Dazzling near-equal whites." The letter 'k' is often confused with the Greek 'kappa' on star charts & in showpiece listings. A fixed pair.
M46/NGC 2438 PUP (13/6)	07 42 -14 49	OC PN	6.1 11.5/16	27' 66"	! A rich uniform cluster of 100+ relatively dim stars with a faint PN projected against its N edge, together set in a "Glorious low-X field." "Beautiful circular cloud of small stars...(with) a feeble nebula on its N verge." The latter "An astonishing & interesting object...annular." "150 faint stars of remarkably uniform brightness...small telescopes show M46 as a sprinkling of stardust." "Fairly rich...a magnificent OC at low X. The

planetary is difficult to locate - oval, non-uniform in brightness & blue-grey in color"
in 4-inch. "Large mass of faint stars." "Annular star-cloud." "Dotted on its N edge"
with an irregular "planetary ring nebula" that's "round & soft." "A noble though
rather loose assemblage of stars...more than filling the field...with power 93...on the
northern verge is an extremely faint PN...the splendid glow of the mass...attentively
gazing...the impression left on the *senses* is that of aweful vastness & bewildering
distances...those bodies bespangling the vastness of space." "A rich OC, beautifully
situated in the glowing stream of the Puppis MW...evokes the impression of a celestial
meadow strewn with fireflies." "Superb." "A densely packed throng bursting at the
seams with...stars." The PN "Exactly round, of a fairly equable light...a very minute
star a little N of center...not brighter in the middle or fading away, but a little velvety
at the edges" as seen in 12-inch & larger instruments. It can be dimly glimpsed in a
5-inch at 50X to 75X, & is fairly obvious in an 8-inch at 80X as a "tiny grey disk"
or ring with a background star shining through the central hole! But this remarkable
combination is only a coincidence, since the cluster & planetary differ greatly in actual
distance (D=5,400LY & 3,000LY, respectively)! Although their separation was not
known in his time, William Herschel deduced *solely from the cluster's visual aspect*
that the planetary has "No connexion with the cluster, which is free of nebulosity."!

M47 PUP (13)	07 37 -14 30	OC	4.4	30'	

! Bright sparse splash of several dozen stars just 1.5 degrees W of M46. "Naked-eye
cluster." "Grand broad group." "Beautiful coarse OC lying in the heart of the Puppis
MW. Many colored stars seen at low Xs. Contains fine DS Struve 1121 (7.9, 7.9, 7")
near center." "Impressive sight, includes several double & colored stars." "DS in a
loose cluster...both bright bluish white...inhabits a very splendid field of large & small
stars dispersed somewhat in a lozenge shape." "Terrific!" "Broad integration of mag.
5-9 suns. Colorful units abound. A majestic zone thruout." "A 6-inch telescope can
resolve just about all of these stellar sapphires as they burn against a velvet black background. The beauty of the scene is further enhanced by surroundings that overflow
with stars." "A coarse scattering of bright stars lies on a dim sheen of fainter one, so
a 10-inch is needed to enjoy its splendor fully." "Beautiful in any scope." D=1,500LY

M93 PUP (10)	07 45 -23 52	OC	6.2	22'	

! "Bright cluster in a rich neighbourhood." "A pretty sight in binoculars." "Arranged
in a wedge shape." "A glorious view. Triangular shaped with many colored stars. A
compact swarm of stellar jewels. Surroundings grand for sweeping at low X." "Bright.
Core has 'V' shape." "Fairly rich" - about 80 members. "Appears knotty with many
vari-colored stars." "This neat group is of a starfish shape...a mass of small stars."
"Smaller but brighter group than M46; the central mass being distinctly triangular or
wedge-shaped with outer branches & scattered sprays of stars...the over-all pattern
suggests a butterfly with open wings." The brighter stars "form a line that zigzags
close to the cluster's center" while in larger apertures the host of fainter ones look like
"Stellar fireflies seemingly flitting about a dim flame." A beautiful sight in anything
from a 2-inch glass at just 25X to a 14-inch at 150X on a dark night! D=3,400LY

NGC 2440 PUP (12)	07 42 -18 13	PN	10.5?/14.3	16"	! "Bright; pale bluish white. With my 64X, like a dull 8th-mag. star: with more X, small, brilliant, undefined, surrounded with a little very faint haziness. In a glorious neighbourhood." "Boxy shape, high surface brightness." "Almost starlike; irregular at high X." "A 20" diameter core with 50" diameter outer envelope; tiny li'l thang!" "Curious bright opal." "Bright bluish PN." A "protoplanetary?" "Visible in scopes as small as 3-inches. A 4- or 6-inch...will show it as a 20" circular patch of turquoise light. With high magnification & large aperture, a faint, strangely rectangular halo can also be seen engulfing the bright center." Visual mag. estimates for this PN are quite discordant, ranging from 9.3 to 11.5. Neat sight in 10-inch, 160X. D=3,500LY
NGC 2477 PUP (5)	07 52 -38 33	OC	5.8	27'	! Perhaps the best of Puppis' amazing hoard of lovely star clusters (dozens of others can be found here which space simply prevents including)! "Large binocular cluster of 300 faint stars, like a loose globular." "Superb, rich with 11th-mag. & fainter stars." "Looks just like a ball of celestial cotton spanning about one Moon diameter through low-X binoculars. Through an RFT...the cluster's fuzziness begins to dissolve into a myriad of faint points of light." "Probably the finest...in Puppis, but not noted by Messier & omitted also from many observing guides because of its position low in the southern Winter sky...a striking group...containing about 300 stars crowded into a 20' field...unusually rich." "Almost globular." "In superficial appearance...the richest of galactic clusters; or perhaps it is the loosest of the globular clusters." Beautiful in a 4-inch at 45X, it's a superb sight in an 8-inch at 80X & the stellar density holds up to make it spectacular even in the largest of amateur instruments! Far S. D=4,000LY
Theta SGE (7)	20 10 +20 55	DS	6.5, 8.5, 7.4 F5IV, G5, K2	12", 84"	"Fine triple" star in rich MW setting with subtle hues. "Yellow-white, ashen, yellow." "Pale topaz, grey, pearly yellow." Delicate in 5-inch at 50X. A-B CPM, C optical.
15 SGE (1)	20 04 +17 04	DS	5.9, 9.1, 6.8, 8.9 G1V, , A2	190", 204", 183"	**Sapphire Star** Guide-beacon to a lovely tinted stellar gem. "Commands another fine group. N, a little preceding, at a few minute's distance a 7th-mag. beautiful sapphire blue" star. Identification is definitely a bit of a problem here! The 7th-mag. comp. to 15 SGE itself does lie 3.4' to the NW - or "N, a little preceding" - but at A2 is not that blue. The author believes that 11 SGE (19 58 +16 47) may be the star referred to by Webb. Although it's 5.3-mag. instead of "7th" & just S instead of N of 15 SGE, its very bluish B9III spectral type (the blue companion of Albireo is B8V) certainly does fit the description of being a "Striking sapphire-colored star."! What do you think? Its tint is perhaps best seen in a 6-inch aperture (not too little - nor too much - light).
M71 SGE (11)	19 54 +18 47	GC?	8.3	7'	! Neat remote-looking globular in rich MW setting! "Dense cluster visible as a misty patch in binoculars & appearing nebulous in small scopes...usually classed as a GC, but some authorities regard it as a rich OC." "Of uncertain type." "Loose globular; looks like an OC." "A beautiful sight...not resolved (in a 4-inch)." A curtain of faint foreground stars is projected against its glow. "Large & dim, hazy to low powers with 3.7-inch, yielding a cloud of faint stars...to higher magnifiers." "Dazzling sweep of

star-clouds!" "A rich compressed MW cluster on the shaft of the arrow." "A gorgeous sight...so easy to resolve...that it was long thought to be a rich OC - one of the 'loosest' GCs...in an exceedingly rich starfield...a glistening backdrop. Resolves clear across its face, yielding a stunning view of a great distant globe of aged suns." "Rich in very faint stars, resolvable, no distinct core." "Oddly shaped; not round! Sparse & dim OC Harvard 20 (10th-mag., 8' dia.) 30' SSW." This "Beautiful low-X field, containing pair, & triple group, all about 8th- or 9th-mag." consists of some 20 stars whose exact "Identification is difficult due to the stellar richness of its surrounding." M71 itself is easily seen even in a 3-inch at 30X. A 10-inch at 100X gives a "magnificent view" & a 14-inch at 150X shows it as a sparkling stellar beehive! Considered to be one of the nearest GCs (or perhaps most distant OCs!): D=8,500LY However, values as great as 18,000LY are given by some astronomers for this cluster with a 'schizoid personality'!

M8/NGC 6530 SGR	18 04 -24 23	DN	5.8		90'x40'	
(21/2)		OC	4.6		15'	

! Lagoon Nebula Wondrous sight! A large floating nebulous patch crossed by great curving dark lane, with a scattered OC to one side - all easily seen in a 2-inch glass. "One of the finest showpieces in the heavens." "Detailed nebulosity with dark lane & cluster." "Perhaps second only to M42 (the Orion Nebula)." "Famous...visible to the naked eye encompassing the star cluster NGC 6530 apparently recently formed from the surrounding gas...an area twice the apparent size of the full Moon...appears milky-white...dark rift down center." "Great lane of obscuring matter crosses center. Adding to the view is NGC 6530, a fairly rich OC. When the air is unsteady, seems suspended among nearby stars" thru binoculars. "An ill-defined nebulosity with dark patches & stars." "Splendid...object. In a large field we find a bright, coarse tripe star, followed by a resolvable (?) luminous mass, including two stars or starry centres, & then by a loose bright cluster enclosed by several stars: a very fine combination." "Myriad's of low-mag. stars, & a few brighter units resembling somewhat the Pleiades, involved in wide wastes of incandescent hydrogen & helium, overflung with dark absorbing patches. A naked-eye wonder." "A large, bright, & very singular body, nearly cut in two by a central dark lane. Various bright & nebulous stars appear involved in the nebula. A magnificent object in long-exposure photos." The "Sprawling Lagoon - one of the sky's brightest, most distinctive & most spectacular emission nebulas...as grand as one can view." "Small telescopes show a glowing cloud of great intricacy sliced in two by a dark lane, or 'lagoon,' of obscuring dust. Through large backyard instruments, the view is truly awe inspiring, with pockets of dark nebulosity sewn between patches of bright clouds...breathtaking in its complexity. On exceptional evenings, it is even possible to catch something of the reddish color that is so prominent in photographs." "The W half of M8 is dominated by two bright stars just 3' apart; the S star is 9 SGR...(which) would appear to be the chief illuminating star of the nebula. Just 3' SW...lies the brightest segment of the nebulosity, a 'figure 8' shaped knot about 30' in size & often called...'The Hourglass'. The E half of M8 contains as its most prominent feature the loose star cluster NGC 6530." "A coarse, rather spherical cluster of bright stars wrapped in a luminous haze." Its shape reminds at least one observer of a "wine goblet." D=5,000LY, which makes the Lagoon at least 60 x 45 LY in size. "One more of the best!" The Trifid Nebula (described below) lies less than 1.5 degrees NW of the Lagoon, & the two "may actually be portions of the same vast nebulous aggregation (since) the derived distances are fairly comparable."!

M17 SGR (20)	18 21 -16 11	DN	6.0	46'x37'

! Horseshoe/Omega/Swan Nebula Multi-named glowing wonder! "A curious... magnificent, arched, & irresolvable nebulosity - in a splendid group of stars." Well seen in a 3-inch glass at 45X, a 6-inch shows it as a long ray, hooked at one end, & crossed by dark lanes. "A cluster associated with nebulosity...35 stars...repays careful & repeated study...most conspicuous portion is the straight bar, which appeared white visually. The sky inside the hook looked particularly dark - a contrast effect?" "Small telescopes show an elongated smudge, appears arched-shaped in larger instruments." "Nice 'check mark' with outlying faint nebulosity." "Bright & wonderful, one of the best for small scopes." "Rivals M42 for detail." "A train of light...in the shape of a spindle." "Looks more like a long checkmark than a horseshoe" & so is sometimes called the "Checkmark Nebula." But "In ordinary 'scopes, more the shape of a swan." "A wonderful extensive nebulosity of the milky kind." "So named because the faintest visible portion of it is curved." "In brightest portion of the MW. Curiously arched. Interesting with low X, but with increased magnification an exquisite object. Legions of nebulae S." "Stunning region of bright nebulosity - a spectacular sight...its soft gossamer glow...small 'scopes reveal an intricate cloud in the shape of an extended number '2'. With increasing aperture the Swan slowly disappears into a huge, glowing, semicircular arc of light...an interstellar spectacle whose magic cannot possibly be conveyed in words or pictures." "Space, the grand theater of astronomical meditation, it here illimitable...so great is the number of stars. Carrying this view into adjoining regions, words & figures necessarily fail, for the powers of mind falter in such vast & awful conceptions." "A very fine object, of arched form, with an interesting group of stars...rich MW region." "The main feature of M17 is the long comet-like streak across the N edge; on the W end a curved 'hook' gives the whole nebula a resemblance to a ghostly figure '2' with the bright streak forming the base. It requires only the slightest use of the imagination to transform this pattern into the graceful figure of a celestial swan floating in a pool of stars." D=5,000LY About 1 degree S lies M18, a small loose cluster of a dozen or so stars (8th-mag., dia. 7'). A "Glorious field in very rich vicinity." "Loose & poor - pretty sight for very small telescopes" only. "One of the most neglected of the M objects." "Raspberries!" Appears on (9) showpiece lists.

M20 SGR (15)	18 03 -23 02	DN	6.3	29'x27'

! Trifid Nebula Although visually much fainter than the nearby Lagoon Nebula & considered "distinctly inferior" to it by most observers, this object still deserves long & careful study. "One of the best! Worth an hour of your time." The "Famous Trifid. A dark-night revelation, even in modest apertures. Bulbous image trisected with dark rifts of interposing opaque cosmic dust-clouds...outlying areas rich in stars in cruciform & other symmetries." "Remarkable dark rifts in bright cloud." "Named for its three-lobed appearance. Very large...interesting nebula...in a beautiful MW field. Dark rifts readily seen in 2.4-inch refractor & non-uniform in brightness in 4-inch refractor. N section has slight greenish tinge (blue in pictures). HN 40 multiple star system at central junction." "Even moderate sized scopes show it as only a diffuse patch of light centered on the DS HN 40." This "Central triple star (7, 8, 10, 11", 5")...appears as a double in small 'scopes & as a multiple system in larger ones." "A beautiful triple... situated precisely on the edge of one of these nebulous masses just where the interior vacancy forks into two channels." "Closely follows a cruciform group. Very curious object; pair with a minute comes 'where the three ways meet, dark rifts through the nebulosity'...nebula imperfectly seen. Grand region." Somewhat of "A challenging object because of its low surface brightness. Fine sight in binoculars - with a 6-inch becomes an exciting object." "Diffuse type, very large & bright (?), & divided into irregular sections by dark lanes. Called a remarkable & curious nebula. With double & multiple stars." "The clouds of the Trifid are rather faint & difficult to discern through small instruments on hazy summer nights...M20 is one of a rare breed that handles magnification well...the soft, gossamer wisps of the Trifid stand out magnificently against a dazzling stellar field" using a WF, high-X eyepiece on an 18-incher. Interesting sight in 3-inch & a fascinating object in an 8-inch on dark nights. About 1.5 degrees NW of M8 & perhaps slightly more distant than it: D=5,500LY The OC M21 (see below) is in the same low-X field, 0.7 degrees to the NE. In a 4-inch RFT at 16X, M20, M21 & M8/NGC 6530 can all be seen together in the same field of view!

M21 SGR (9)	18 05 -22 30	OC	5.9	13'	

Lying just 40' NE of M20 is a bright stellar clan of some 60 stars "In a lucid region." "Very fine & impressive - rather small & compact diamond shape." "Coarse cluster of telescopic stars in a rich gathering galaxy region." "Outstanding in binoculars. If it were situated alone in the sky, M21 would still be a fine showpiece object. However, the added beauty of its magnificent surroundings moves this OC up the ranks to a 'true stellar gem'." "Beautiful, in a nest of nebulae." "Coarse cluster of telescopic stars." "A fairly compact group highlighted by about six 8th-magnitude stars in a tight knot, surrounded by several dozen more scattered members." "Unimpressive at first glance; close scrutiny yields the smaller (dim) members which flesh this beast out - very, very pretty!" "M21 looks good in just about any amateur telescope" & together with M20 "Forms a sparkling pair of celestial celebrities." Nice in 3-inch glass at 30X - appears at its glittering best in an 8- to 10-inch scope at 60X to 80X. D=4,000LY

M22 SGR (20)	18 36 -23 54	GC	5.1	24'	

! M13 Rival Along with M5 (SER), this magnificent stellar beehive gives the famed Hercules Cluster a real "run for its money"! Indeed a "Rich GC & one of the finest in the entire heavens - ranked third only to Omega CEN & 47 TUC...noticeably elliptical outline...3-inch scopes begin to resolve its outer regions - larger apertures show the brightest stars to be reddish." It was actually the "First globular known (1665)...as impressive as M13. Partially resolved (in 4-inch) but not its center, which remains a solid glow." "Beautiful large globular, extremely dense." "Spectacular from southern latitude." A sparkling fuzzy ball in a 3-inch glass at 45X, while 75X breaks it up into individual pinpoints of light. Easily resolved to center in 10- to 14-inch scopes, which show many of the stars to be orange or reddish in tint! "Beautiful bright cluster, very interesting from the visibility of its components, largest (brightest) 10th- & 11th-mag, which makes it a valuable object for common telescopes, & a clue to the structure of many more distant or difficult nebulae. John Herschel makes all the stars of two sizes, '10-11th-, & 15th-mag., as if one shell over another' & thinks the larger ones ruddy." "Another colossus...1/2 degree in extent. Larger & brighter than M13 HER, tho less condensed." "Superb." "Called the finest cluster after M13 visible from N latitudes. Bright & very fine, composed of 10th-mag. & fainter stars, so compact that the object is visible to the naked eye." "Rivaling M13. At 200X in 8-inch a great big ball of sparklers shimmering with a feeble light across the field." "Unusual GC; center so condensed that it looks like an 8th-mag. star with a hazy surround." Yet as prominent as it appears, M22 is obscured by interstellar dust - without which it "Would be *5 times as bright* as we now see it!"!!! "Its richness & easy resolution will astonish the small scope user. The milky glow alive with tiny, individual images, will leave any tele-scopist impressed." "One of the sky's greatest unsung sights. Take a careful look at the cluster. Do you notice anything peculiar about its shape? Unlike most globulars, M22 in not perfectly round...(a) NE-SW bulge is noticeable" - visual proof of leisurely rotation on its axis! "Consists of very minute & thickly condensed particles (photons!) of light." "Wonderful...one of the easiest of the globulars to resolve." As mentioned under M13 above, the view of a big globular in an observatory-class scope is simply unbelievable: "What gorgeous sights these clusters are when there is enough light to activate one's color vision! The center of a GC is a jewel box of topazes, garnets, rubies, & the odd amethyst - all in profusion...a never-to-be-forgotten personal observ-ing experience." "You can see the colours of the stars down to about the 15th-mag. In addition to the densely packed mass of individually resolved stars...the whole cluster shows a granular background - it's rather like looking into a bowl of sugar! There must be perhaps 200 red giant stars...& you can pick them out at sight, even in the very middle of the cluster." "It was as if a globe had been filled with moonlight & hung before them in a net woven of the glint of frosty stars." This is what awaits you on a personal visit to a major observatory public viewing night! At least 500,000 stars right before your eyes! One of nearest GCs: D=10,000LY (less than half that of M13!)

M23 SGR   17 57 -19 01   OC   5.5   27'
(15)

! Big, uniform & fairly rich cluster. Most striking in medium apertures at low X - the stars thin out too much in really large amateur & observatory-class telescopes. "Rich widespread cluster...elongated in shape...remarkably uniform appearance." Truly "A glorious sight - lies in grand star field...brightest stars form pattern resembling a bat in flight." "Beautiful, chains & loops, use low X." "Grand low-power field. Announced by increasing number." "Bright, loose cluster." Here's "A blazing wilderness of starry jewels!" "Not terribly well detached from the crowded (MW) background." Rated as "Comparable to the following cluster (NGC 869, W member of the Double Cluster) in

PER." "An elegant sprinkling of telescopic stars over the whole field." "Brilliant. Appears as grains of glistening sand on the black velvet of space." "A beautiful field of glittering stars." Pretty in a 6-inch at 50X, but "The most pleasing view of M23 (is) to be obtained with about 45X on a 10-inch f/6 reflector." Striking in an 8-inch at 60X, less so in a 13-inch at 145X. Over 150 suns reside in this clan. D=2,100LY

M24/NGC 6603 SGR	18 18 -18 25	MW	4.5	60'x120'	
(13/0)		OC	11?	5'	

! Small Sagittarius Star Cloud "Rich & extensive MW starfield." "2.2 degree x 1.3 degree extent! Talk about having your head in the stars! Naked eye, binoculars, telescope...use 'em all!" "Huge, rich at low X." "Rich star cloud; contains OC NGC 6603." "Designation (M24) has been applied to two different objects, one within the other. Not a true galactic cluster - a small detached portion of the MW easily seen with the naked eye & a fine sight in binoculars & RFTs. NGC 6603 *is* a true galactic cluster - 'A gathering spot with much star dust'." Containing perhaps as many as 100 stars, it's a "Beautiful compact cluster in the NE corner of M24" itself. "Beautiful if somewhat subtle" in the telescope. "A faint but very rich group...but not easy to detect in any aperture smaller than 8-inches." "A tight pack" - so much so that "Its misty appearance has led more than a few observers to conclude erroneously that it is a GC." "Globular cluster...in superb field."! Suspended midway along M24's right side is **Barnard 92**, one of the most prominent dark nebulae in the sky. It's "Extremely opaque" & "Impossible to miss in a small telescope."! "Seen in silhouette against the NW corner of the star cloud...relatively easy to see, even through light-polluted suburban skies, as an oval void measuring about 12'x6'." "The most stunning dark nebula in small telescopes...floats in front of the bright, rich starcloud." M24 itself is truly "A region of indescribable richness." "Magnificent...visible to the unaided eye as a kind of protuberance of the Galaxy." "A grand sight in any low-X telescopic field ...spectacular sight in binoculars." "A beautiful stellar field." A "Giant cluster!" "Indeed one of the finest & richest in the heavens for wide-angle scopes." Star cloud, cluster & nebula seen together in same view in 6-inch RFT at 24X - an amazing sight! D=16,000LY for both M24 & NGC 6603. (There's also a large or "Great Sagittarius Star Cloud" which marks the direction to the center of our MW Galaxy, located just off Gamma SGR in the 'spout' of the striking 'Teapot' asterism. This is a fantastic part of the heavens to sweep with binoculars & WF telescopes on dark transparent nights!)

M25 SGR	18 32 -19 15	OC	4.6	32'
(11)				

! A large splashy cluster of some 50 suns that's "Unusual in containing, as one of its members, the bright classical Cepheid variable star called U SGR" (mag. range 6.3 to 7.1 in 6.7-day period) near its center. "Coarse & brilliant." "Somewhat diffuse but strong." "Bright but sparse cluster." "A superb sight for small apertures, with many colored stars." "Superb scattering, includes several colored stars, dark pools." "Loose cluster of large & small stars...thickly strewn in the S where a pretty knot of minute glimmers occupies the centre, with much star-dust around." "Spectacular...showpiece for binoculars & RFTs." "Commanding OC." Grand sight in 6-inch RFT at 24X! Surprisingly, not included in the NGC - but did make it into the IC. D=2,000LY

M55 SGR (14)	19 40 -30 58	GC	7.0	19'

! Nice ball of stars at fairly low altitude for most N observers. "Exceptionally loose. S DEC limits status. A gem of a globular - would rank among the finest if placed higher." "Bright, loose globular." Requires a really first-class night for a good view, at which time it resembles M13. Hazy sparkling patch in a 3- or 4-inch. "Nebulous looking...small scopes resolve individual stars but show little central condensation." "The 4-inch did not resolve the cluster nor reveal grainy texture. Center contains a bright star." "Very large, loosely compressed globular, easily resolved" in 8-inch & bigger apertures. "Large & distinct though faint; resolved but grainily" in an 8-inch. "The least concentrated globular in Messier's catalog. Low X (on an 8- to 10-inch) will resolve many individual points of light sprinkled across the nucleus...giving the impression of a sparsely populated cluster." "Plainly stellar...a huge agglomeration of stars uniformly distributed & immersed in a pale nebulosity (of unresolved faint background stars)...should be admirable in the S hemisphere; for us it is a little pale." In a 14-inch at 150X on a dark, steady night this superb glittering orb looks very much like a fainter version of M13 & M22. Among the closer globulars: D=16,000LY

NGC 6818 SGR (12)	19 44 -14 09	PN	9.9/13.0	22"x15"

(Please see Page 96)

! **Little Gem Nebula** A small but exquisitely-tinted planetary - in common with many other members of its class, starlike in 2.4- to 4-inch glasses at low X but reveals a tiny disk when more highly magnified. "Blue, like star out of focus." "Small, high surface brightness, bluish." "Like a monster fish-eye" in 10-inch! "Pale blue PN...four stars form a square about it." "Greenish disc." "A uniform disk of turquoise light...'scopes of 8-inches & larger at higher magnifications add some personality...by hinting at the cloud's true ringlike structure." Also described as resembling the "CBS Eye" in big backyard instruments! Curiously, one noted observer of bygone days stated: "Beyond the reach of a 6-inch even in clear mountain air, but easy with 12-inch."!? The visual mag. of this object seems uncertain, values from 9.5 to 11.1 being given in various lists & catalogs. The brightness of its faint nucleus is also not pinned down, being listed as anywhere from 13th- to 15th-mag. "Annular; NGC 6822 **Barnard's Dwarf Galaxy** 3/4 degree S." This famous but extremely faint 16'x14' smudge of light lies 45' SSE of the planetary, within the same wide eyepiece field. "Very large, very low surface brightness." "Requires transparent skies." "Actually somewhat easier to detect than the Veil Nebula in CYG." (It was discovered visually by Barnard with a 5-inch glass.) Like its neighbor, it too has an uncertain visual mag. - between 8.8 & 11.0, according to different sources! On (6) showpiece lists. "Elusive!" NGC 6818 D=5,000LY

Alpha SCO (15)	16 29 -26 26	DS	0.9-1.8, 5.4 M2I, B3V	3"

! **Antares** "Rival of Mars." A beautiful fiery-red supergiant with vivid emerald-green companion! Needs at least a 5-inch most nights & tough even in a 6- or 8-inch unless seeing (image steadiness) is good. The pair is a lovely sight in 10-inch & larger telescopes. In a 13-inch refractor at 190X, it is perhaps the most striking of all doubles! Even when Antares is not resolved, the small star can be recognized as a green tinge to the W side of the red primary, as pointed out in the best description of this object ever given - by Webb nearly a century & a half ago: "This great star...is a grand telescopic object. Its tint, however, is not uniform; to me the disc appears yellow, with flashes of deep crimson alternating with a less proportion of fine green, the latter mixture perhaps accounted for by the 7th-mag....star near enough to be usually involved in the flaming rays of the principal, forming an atmospheric, rather than optical test...a curious proof of its independent, not contrasted green light, when it emerged, in 1856, from behind the dark limb of the Moon." An important point! While many - perhaps most - DS tints are the result of contrast effects, some of the color differences seen *are real* (as hiding one component behind an occulting bar in the eyepiece will verify)! "Red supergiant with blue companion, so close it requires...steadiness of atmospheric conditions to be visible against the primary's glare." Called a "brilliant orange-red", the primary star looks tiger-lily orange with definite red tinge to the author! "Burning Antares." "Saffron rose." "Orange & green." "Fire-bright prodigy...vivid emerald companion." "Notoriously difficult when...low in the sky, but easily seen with 3-inch when high...easier in strong twilight or moonlight than against a dark sky." Under good conditions, "It appears quite plainly in a 6-inch telescope as a little spark of glittering emerald, almost drowned out in the blazing ruddy light of giant Antares." Other tints assigned to the small star are "pale green", "vivid green", "verdant", "purple" & "very blue"! Some sources list this object as being 6.5-mag. - 6.0 does perhaps better match what the eye sees than the 5.4 listed here. Antares itself is "A superstar, a giant among giants" - a pulsating semiregular variable with a diameter of 700 x that of our Sun! The two stars share a CPM & are a long-period binary having an estimated 900-year orbit. "Definitely forms a true physical pair." D=520LY

Beta SCO (19)	16 05 -19 48	DS	2.6, 4.9 B0V, B2V	14"	

! Graffias A slightly fainter version of famed Mizar (Zeta UMA), beautiful in all size instruments! "One of the finest bright DSs in the sky for small telescopes." Lovely even in a 2-inch glass at 25X. "Striking...blue-white stars." "Intense, colorful." "Pale yellow, greenish?" "Both bluish." "Pale white, lilac tinge." "One of the most enticing bright doubles." "Very fine." Both stars are spectroscopic binaries & the primary has an ultra-close 9.5-mag. companion seen only in very large scopes. CPM. D=600LY

Nu SCO (14)	16 12 -19 28	DS	4.3, 6.8 B3V	1.2"	
		DS	6.4, 7.8 A0IV	2.3"	

! A subtly-tinted, tight double-double system for medium to large apertures. A 3-inch usually shows only a wide double 41" apart but sometimes splits one of the close pairs, while all four stars can be seen in a 4-inch at 100X. A 5-inch at the same magnification in good seeing easily shows Nu quadruple & color contrasts become evident in 6- & 8-inch glasses. "Quadruple star similar to the famous Double-Double...small scope shows it as a wide double with blue-white components." "Very difficult." "All cleanly split at 322X" in 8-inch, but much less will do it! "Perhaps the most beautiful quadruple group in the heavens, from the narrow limits within which the brilliant objects composing it are crowded." Smyth missed both close pairs in his 6-inch glass, calling Nu "A neat DS. Bright white & pale lilac."! Barns also logged only two stars in his 10-inch, but commented that "Both doubles are double."?? Tints *can* be seen here but pinning them down is a bit tricky. CPM - all one vast system! D=400LY

Xi SCO/Struve 1999 (16/1)	16 04 -11 22	DS	4.8, 5.1, 7.3 F5IV, F5IV	0.3", 8"	! Unusual multiple system - a close triple combined with a wider but fainter pair - that currently appears as a double-double. Like a wider version of Nu SCO in most telescopes. Xi itself consists of a very close binary with 46-year period - presently only a blended or elongated image "with a bright golden tint" in most instruments, together with a more distant third star. "Celebrated multiple star...at its widest in 1976 (1.2") were divisible in apertures of 4-inches but when closest in 1997 (0.2") were impossible to split in amateur-sized scopes." Some 280" to the S lies Struve 1999, looking "Like auto headlights" & together with Xi creating a "Beautiful field." "Deep yellow pair." "Blue or purple, yellowish white." "One of the most interesting of the multiple star systems." All five suns traveling through space together as a family (CPM)! D=80LY
		DS	7.4, 8.1 G8V, K5III	12"	
Mu-1/Mu-2 SCO (1)	16 52 -38 03	DS	3.0, 3.6 B2V+B, B2IV	346"	* A prime example of a wide matched, naked-eye/binocular/telescope pair, similar to Theta TAU in the Hyades. "The Inseparable Ones." "May be considered a physical pair" - CPM. (Others in this area are Zeta SCO, Lambda & Upsilon SCO - the famed 'Stingers' in the Scorpion's tail - & Beta in SGR; the sky's filled with them!) Like all of its kind - a neat sight in WF scopes. Blue-white gems in a 4-inch RFT at 16X. Primary itself is an eclipsing/spectroscopic binary (note dual spectral type). D=520LY
M4 SCO (17)	16 24 -26 32	GC	5.9	26'	! **Oblate Globular** Big, softly shining globular swarm visible in a 2-inch glass & a superb sight in all scopes! "Bright globular near Antares." "Large, rather dim, resolvable, followed by a vacant starless space." "A beautiful object...stubby band of stars in interior...conspicuous globular." "In a 4-inch telescope, individual stars are resolved & there's a noticeable bar of stars across the cluster's center." "Nicely resolved, with 'bar' through center." "Dim but interesting star-cloud." "Right next door to Antares...only a granular disk to a 3-inch." "Magnificent lovely big glow." "Like M13 but less so." "One of the greatest globulars in the firmament." "Compact mass of very small stars, elongated vertically, & has the aspect of a large, pale, granulated nebula, running up to a blaze in the centre." "One of the largest objects of its type, & also one of the nearest (D=7,000LY!)...probably the easiest of all the bright globulars to locate (just 1.3 degrees W of Antares). The central bar makes the cluster appear to be quite oblate in small instruments." (The "oblateness" referred to in the descriptions of many globulars results from slow rotation about their axes; OCs aren't concentrated enough in most cases to notice the effect of their spinning.) "One of the sky's easiest globulars to resolve into individual stars...with a unique 'starbar' bisecting the cluster from N to S." Centrally resolved in 6-inch & larger apertures, which show many apparent star-chains & hints of color. "A huge globular...dynamic object... mighty impressive." "Exceptional." An awesome sight in 10- to 14-inch instruments!
M6 SCO (16)	17 40 -32 13	OC	4.2	25'	! **Butterfly Cluster** "A most beautiful OC...'like a butterfly with open wings'." "Stars radiate from the center, suggesting the appearance of an open flower...certainly one of the finest objects in the sky." "Impressive...stars arranged in radiating chains." "One of the finest sights in the heavens - a grand object." "Famous for butterfly shape." A

large bright scattered clan of some 80 suns requiring at least a 1-degree field for full effect. With nearby M7, "Two superb naked-eye star clusters" - "A pair of beautiful OCs." "Nicknamed the Butterfly Cluster for its remarkable resemblance to one of these graceful insects in flight. Unfortunately, long-focal-length telescopes have fields of view too restricted to encompass the full glory of M6. Binoculars & RFTs, on the other hand, display a wondrous sight. The brightest star...is a blazing orange light-house known as BM SCO." (This K-type giant is located in the NW 'wing tip' & pulsates between mags. 6.8 & 8.7 over an 850-day period.) "Stars...very dispersed & arranged in a remarkable pattern...three starry avenues leading to a large square." "A completely charming group...at its best in a good 6- or 8-inch glass with wide-angle oculars." Stunning sight even in a 3-inch at 30X. Seen in binoculars & WF scopes, this showpiece definitely "Lives up to its Butterfly Cluster nickname"! D=1,400LY

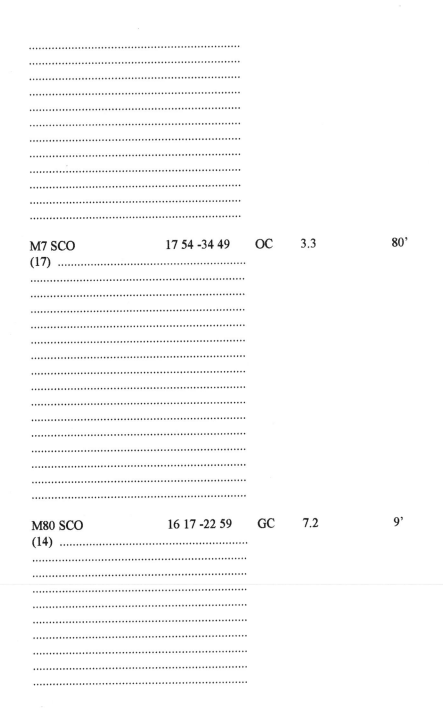

M7 SCO (17)	17 54 -34 49	OC	3.3	80'	

! Sprawling, radiant OC of 80 stellar jewels in the Scorpius MW! "Large, brilliant... visible to the naked eye...brightest stars appear to be arranged in chains." "Beautiful grouping...many yellow & orange stars near center." "Easy naked-eye, gorgeous in any scope." "Excellent in binoculars or RFT." "Looks like M6 but twice as large; each's brightest star is deep orange." "Brilliant, scattered mass of bright stars." "One of Pickering's finest." "Many of the stars are arranged in lines at right angles." "Vivid in field glasses, M7 bursts into an exceptionally beautiful array of stars." "Magnificent family of stars...7X50 wide-angle binoculars create a three-dimensional effect, many of the brighter members appearing to float in front of a field strewn with fainter points of light. Colors abound in the stars of M7, several exhibiting subtle hues of yellow & blue." Like M6, a truly stunning sight in a 3-inch at 30X! "One degree diameter!" D=800LY The dim GC NGC 6453 lies 20' NW of the second brightest cluster star, appearing as "An 11th-mag. fuzzy spot, about 1' in diameter" (some references give 10th-mag. & size as 4'). "Not exactly impressive figures, but it adds to the thrill of the hunt." And the faint OC Harvard 18 lies in the same wide field as M7, 45' to the SE (size 15', 10th-mag., 80 stars). Both of these objects are elusive, & need a 6-inch or larger telescope - plus really dark skies - to be identified with certainty.

M80 SCO (14)	16 17 -22 59	GC	7.2	9'	

! **Herschel's Delight** "The richest & most condensed mass of stars which the firmament can offer to the contemplation of astronomers."! So wrote Sir William, leading to one of this author's earliest disappointments in deep-sky observing! Little did a novice young stargazer realize that this account was made using Herschel's 48-inch reflector! A "Very compressed globular." "Strongly resembles an 8th-mag. comet in small telescopes. Splendid object." "Appearing like the fuzzy head of a comet." "Tiny, partial resolution at high power." "Like a comet; in a beautiful field, halfway between Alpha & Beta...on the W edge of a vast starless opening 4 degrees broad (Herschel's famed "Hole in the Heavens"). Nearly central, is the strange variable, T SCO, which, 1860, between May 18 & 21, had blazed out to 7th-mag., extinguishing apparently the cluster, had almost faded by June 16, & has never distinctly reappeared." (This star is now classified as a nova.) "A round bright nebula in ordinary 'scopes." "Brilliant GC with a blazing center." "Downright impressive." "Very beautiful." "A compressed mass of stellar points reminding early observers of a comet." "Set in a striking star-field." "In ideal sky conditions, an 8-inch will just show some of the 14th-mag. points of light that make up this huge globe of stars...high magnification is a must, however, as the cluster's high stellar density hampers resolution at less than about 150X." A "Very strong central condensation resolved into individual stars at 166X (in 8-inch); beautiful!" "From the blazing centre & attenuated disc, it has a very cometary aspect. Splendid conglomerate." "Some hint of resolution is evident in a 6- or 8-inch 'scope, but the true splendor of M80 is reserved for the fortunate few who have access to great telescopes." A fascinating sight in a 13-inch Fitz-Clark refractor at 270X! And after the passage of many years, the author did finally have the opportunity of seeing M80 *just as Herschel described it* - through the eyepiece of a 30-inch Brashear refractor at a magnification of 600X! A truly amazing & unforgettable spectacle! D=27,000LY

NGC 6231 SCO (6)	16 54 -41 48	OC	2.6		15'

! "Large naked-eye cluster of 120 or so suns in rich area of the MW. The brightest stars of the group are 6th-mag. & give the impression of being a mini-Pleiades. The blue supergiant Zeta-1 SCO, an outlying member...is connected to a larger, scattered cluster of faint stars...called Harvard 12 (mag. 8.5, 40' dia., 200 stars!), lies 1 degree to N with a chain of stars linking the two." "Beautiful, bright glorious cluster" of O-type supergiant suns! "Superb low-X field, bright stars in center." "Dazzling...brilliance in low-X fields will startle you. Looks like luminescent buckshot, sprayed onto the velvet black background of interstellar space. If (placed) as close as the Pleiades, it would outshine it by a factor of 50" & its brightest stars "would rival even blazing Sirius."! "Gorgeous star cluster." "A 'must see' cluster." "Brilliant 5th-mag. star overlays 100-plus 10th-mag. members. OC H12 1.5 degrees to NNE...add the Zeta SCO crowd & the area makes a terrific binocular field." "An 8-inch telescope at low X...displays about a quarter of its jewel-like stars...the remaining fainter cluster members creating a triangular wedge of celestial mist. Larger apertures & higher magnifications blow some of the cloudiness away, revealing even more stars." This magnificent stellar jewel-box is unfortunately often overlooked due to its low altitude above the horizon. "A striking & impressive object...would require only a more favorable position in the N sky to make it one of the most famous objects in the heavens...glittering diamonds displayed on black velvet...very little color in this cluster; all the brighter stars appear brilliantly white...many bright stars, pairs & triplets." "Resembles a miniature edition of the Pleiades, with a central knot of 7 or 8 bright stars" in small glass. "An amazing sight" in 5- or 6-inch at 50X! "One of the finest." Don't miss it! D=6,000LY In the same wide eyepiece field, 1/2 degree to the S, is the "Wide naked-eye star pair...Zeta-1 & -2 SCO, about 6.8' (408") apart, with a very noticeable color contrast" (mags. 4.8 & 3.6, spectral types a blue-white B & an orange K, respectively). Sweeping this entire area with binoculars & RFTs on a dark (Moonless), transparent night, adrift in the rich MW starclouds of SCO, is a memorable stargazing experience! H12 itself on (4) lists.

NGC 6302 SCO (5)	17 14 -37 06	PN/DN?	9.7/10	2'x1'	**Bug Nebula** Strange-looking object with uncertain identity. "Much flattened Figure-8 shape; possible unusual planetary." "Very bright elliptical gaseous nebula." "Curious bi-polar object." "Elongated, cut by dark lane." *Sky Cat 2000.0* lists it under PNs, but adds the note "H II region (emission nebula)?" It's quite big for a planetary (although some sources list it as small as 45" in diameter??). An easy catch in a 6-inch glass at 50X, yet it's seldom found on observing lists! In very rich MW starfield. D=1,900LY
R SCL (1)	01 27 -32 33	SS	5.9-8.8 N7.7/C6	---	A ruddy Mira-type pulsator & carbon star with 370-day period. "Deep red variable." One of the reddest stars in *Sky Cat 2000.0*. "Deep orange." Tint obvious in a 4-inch glass, but is perhaps most stunning in an 8-inch reflector at median brightness. Some sources give the visual mag. range as 5.0 to 9.0, with average light around 6th-mag.
NGC 55 SCL (8)	00 15 -39 11	SG	7.9	32'x6'	! "SG seen nearly edge-on, similar in size & shape to NGC 253 (below), though not quite as bright." An amazing 1/2 degree in length! "Huge, irregular edge-on galaxy with faint E section." "Huge barred spiral, edge-on...mottled." "One of the showpiece galaxies of the Autumn sky, goes ignored & unappreciated...because of its extreme S DEC & great distance from bright reference stars...causing most observers to miss out on its true splendor. When...highest in the sky, even an 8X50 finderscope will reveal its long, needle-thin disk...an amazingly intricate object riddled with many bright & dark patches" in medium to large apertures. "One of the outstanding galaxies of the S heavens." Due to light-loss from atmospheric absorption & haze at its low altitude, needs at least 6-inch aperture to fully appreciate. A member (along with NGC 253) of the "Sculptor Group" galaxy cluster. "Incredible!" "Breathtaking." D=7,000,000LY
NGC 253 SCL (17)	00 48 -25 17	SG	7.1	25'x7'	! **Sculptor Galaxy** Another big & even brighter spiral that looks somewhat like a smaller version of the great Andromeda Galaxy (M31) itself! "Very large & bright but at low altitude", needing a first-class night for an optimum view. "Seen nearly edge-on, appearing cigar-shaped...nearly 1/2 degree long...requires at least a 4-inch aperture to distinguish the central bulge & tightly wound arms." "Elongated, mottled dark lane W of nucleus." "Conspicuous extended nebula." "Very bright large nebula." "One of the finest." "Very fine large edgewise spiral." "Oblique view with axis oriented NE-SW; great binocular object; mottling visible at greater than 200X on 8-inch. "One of the largest, brightest, most detailed in the sky...large silvery needle of light. Bettered only by the Magellanic Clouds (MW's two satellite galaxies) & M31 as best galaxy for small scopes. Obviously looks like a galaxy." Actually the 5th-best of its class in the heavens - counting the MW itself as 1st! "Very large; the next brightest of the spirals after M31." "Southern sky's answer to the Andromeda Galaxy. Impressive sight. Tipped dinner plate easily visible in binoculars." "Singular object. A long narrow nebula...of a pale milky tint." (The word 'nebula' was used by early observers for both real nebulosities & galaxies, before the true nature of the latter became known.) "Well worth viewing." "An exciting galaxy to view through just about any instrument. Even from N observing sites, NGC 253 stages an impressive performance. Its oval nucleus

& long, slender galactic disk shine brightly in a 6- to 8-inch...while a 10-inch or larger will disclose the entire span of this magnificent object." "A superb object." "Probably the most easily observed spiral with the exception of...M31. Owing to the S DEC, it should be observed when near the meridian" - excellent advice for viewing all celestial objects, but especially so for those at low elevations. "Spectacular." D=7,500,000LY

NGC 288 SCL (4)	00 53 -26 35	GC	8.1	14'	

Just 1.5 degrees SE of NGC 253 is a large, but rather dim GC that's often overlooked by observers. "Compressed glitter - easily resolved at 100X" on an 8-inch. "Speckled haze. Impressive globular. Loose, resolves easily." "Loose structured." An "Open type globular." "Time for a break. In the midst of all the galaxies strewn across the Autumn sky is this globular cluster, a refreshing deep-sky oasis...however, NGC 288 is not as easy to spy as many other globulars. Its S DEC, coupled with its large apparent diameter & inherent dimness, can challenge even seasoned observers." Although it's as bright as mag. 7.2 according to some sources & visible in small scopes, this is really a medium-aperture object. Glitters to its core in a 12- to 14-inch on dark, steady night.

M11 SCT (21)	18 51 -06 16	OC	5.8	14'	

! Wild Duck Cluster The finest OC in the sky N of -45 DEC for larger instruments (which typically spread out these clans unduly) & a spectacular sight in all size scopes! A very rich & compact stellar swarm. "Perhaps the best OC." A "Noble fan-shaped cluster...at the upper edge of the broad luminous cloud which marks Scutum. William Herschel...divided it into 5 or 6 groups, noted independently with 5.5-inch. An 8th-mag. star is a little within its apex; an open 8th-mag. pair S following beyond it. Just visible to the naked eye...'dark structures' plainly visible." "A real sparkler!" "The richest of all bright OCs. Beautiful sight for Summer stargazers...500 stars...unforgettable...striking...in 8-inch at 50X...resolvable in any scope. An observer inside the cluster would see several hundred 1st-mag. stars in the night sky."! "Magnificent." "A splendid cluster. This object, which somewhat resembles a flight of wild ducks in shape, is a gathering of minute stars, with a prominent 8th-mag. (one) in the middle." "Very very rich for OC...truly a beauty!" "Semi-globular, in a star cloud visible to the unaided eye in a dark sky." "In 6-inch & larger 'scopes on a good night, this cluster can be quite awesome...arguably the nicest of all the OCs." An "Exceedingly beautiful aggregation of small stars." "A grand fan-shaped cluster, with bright star at its apex. Dark structures to S." A "Semi-globular OC." "Famed for its high density...a stellar traffic jam. Thru binoculars looks more like an unresolved GC than an OC." "One of the richest & most compact OCs - 600+ stars - one of finest views in the heavens for small apertures." "Arrowhead shape, rich in faint stars." "A showpiece OC of about 200 stars, noticeable fan shape...100X breaks it up into a sparkling field of stardust." Discovered in 1681 by Kirch, who saw just a "small, obscure spot with a star shining through."! "Sure enough, if you look at M11 through a small telescope it reveals only a long 8th-mag. sun set amid a nebulous glow. Larger instruments burst the cluster into a crowd of fainter stars...in a blunt V pattern." "Exceptionally fine...lying on the N edge of the prominent Scutum Star Cloud (see below)...outstanding. In binoculars or a low-X telescope it at first resembles a globular, but with increasing magnification the stars begin to draw apart, finally revealing M11 as a rich swarm of glittering star points, somewhat triangular in shape with one brighter star near the center, but no real central nucleus." "A prodigious cluster of very small (i.e. faint) stars, forming a large white cloud." "A magnificent pile of innumerable stars." "A carpet of sparkling suns to the very center with outlyers swarming on all sides. A...10-inch shows hundreds of glittering star points all over the field of view." Beautifully resolved in a 4-inch glass, simply awesome in a 13-inch! "A big eerie apparition emerging from the deep." An impression of dark lanes crossing cluster strong in all apertures. One of the very few of its class rich enough to remain striking in a 30-inch aperture. "M11 is considerably nearer to us than the Scutum Cloud & is not directly involved in it." D=5,500LY Strangely, M11 was *not* included by Pickering in his famous showpiece list?! (It *was* entered on his survey's observing roster, where it appeared under globular clusters.)

NGC 6712 (8)	18 53 -08 42	GC	8.2	7'	

A small misty-looking globular, overshadowed by M11 & seldom observed. "Beautiful resolvable nebula." "A beautiful globular star cluster." An "Irregular outline, requires high X to resolve." Some sources give apparent size as just 3' of arc?? "A vague yet delicate, galaxy-like glow behind...stars of rich MW." "Formless glow with brighter center; just barely resolved at 200X; 14th-mag. stars over diffuse background glow" in 8-inch. Definitely needs big aperture to enjoy! "Look for IC 1295 in field" (14th-mag. PN, 80"x60" in size, 20' ESE). "If you're feeling really lucky, try it!" D=25,000LY

Milky Way SCT (1)	18 40 -06 00	MW	---	---	

Scutum Star Cloud/Gem of the Milky Way One of the brightest & most spectacular sections of our MW Galaxy, overflowing the 12 degree x 9 degree boundaries of SCT! "The entire constellation might be thought of as one big naked-eye deep-sky object." A binocular & RFT wonderland! "The sky's largest deep-sky wonder - most detailed galaxy of all goes virtually unnoticed." "Wondrous firmamental clusters...many astral splashes in this crowded district of the Galaxy." "Downtown Milky Way"! And here - as with the great big billowy starclouds of SCO, SGR & CYG - watch for an amazing '3-D' effect that can occur without warning: as the eye-brain combination makes the association that the fainter stars you're seeing in the Cloud are farther away than are the brighter ones - *that you're actually looking at layer upon layer of stars* - the MW *can suddenly jump right out of the sky at you* in a striking illusion of depth-perception!

Delta SER (18)	15 35 +10 32	DS	4.2, 5.2 F0IV, F0IV	4"	

! Stunning, neatly-paired duo with white or off-white hues. Nicely split even in small glasses, yet close enough to retain their charm in big scopes. "White, bluish white." "Yellow-white & ashen." "White, greenish yellow." "Both blue-white." "Beautiful." "Strikingly placed." "An elegant DS...bright white, bluish white; but under the very best vision, both have a bluish tinge, which, in such a pair, is rather against the theory of contrast." "Beautiful blue & orange, appears wider." Lovely in a 3.5-inch at 40X, where the two stars appear to be just touching each other. An ultra-long-period (over 3,000 years!) binary, very slowly opening from 2.7" in the mid-1800's. D=85LY

Theta SER (15)	18 56 +04 12	DS	4.5, 5.4 A5V, A5V	22"

! "One of the most attractive DSs in the sky for a very small telescope, & is generally resolvable in good binoculars." "Imperial pair in regal setting!" "Noble pair, in a very fine field." "An elegant pair of white stars." "Bright, wide pair of yellow-white stars." "An attractive yellow pair." "Clear sparkling white." "Superb pair." "Exquisite." "Fine white pair - seems brighter." "Neat DS. Pale yellow, golden yellow...easy." "Both blue-white; worthwhile pair at the tip of the Serpent's tail!" A neat sight even in a 2-inch glass at 25X & a very stunning object in a 6-inch at 50X. D=130LY

M5 SER (21)	15 19 +02 05	GC	5.8	17'

! **M13 Rival** A magnificent ball of stars considered "But slightly inferior to that in Hercules (M13)"! A "Beautiful assemblage of minute stars...greatly compressed in centre." "This superb object is a noble mass, refreshing to the senses after searching for faint objects." "Has a bright central blaze & outliers in many directions." "Very beautiful object." "A starry blizzard." "Like a snowball." "Naked-eye beauty! Easily resolved at 100X" with an 8-inch. "Very bright, superb GC...very compressed." A "Glowing compress." "One of the finest globulars." "Spectacular, well-resolved to small core, straw & blue" tinted stars. "Suggests a spider...core has triangular shape... a very beautiful sight." "Stunning GC that ranks with M22 in SGR & M13 in HER." Over 500,000 stars! Truly mind-blowing in large amateur & observatory-class scopes: its "Myriads of glistening points shimmering over a soft background of starry mist, illumined as though by moonlight, formed a striking contrast to the darkness of the night sky. For a few blissful moments, during which the watcher gazed on this scene, it suggested a veritable glimpse of the heavens beyond." "In that first stunning view it seemed as if the fireflies of a thousand summer nights had been gathered here, frozen forever in time & suspended among the stars." (Both observations made with 40-inch telescopes.) "Regarded...second only to famous M13...scopes of 4-inches or more reveal a brilliant condensed center & mottled outer regions with apparent chains of stars radiating outwards." "One of the great show objects of the Summer sky, ranking with M13 in HER & M3 in CVN as one of the three finest globulars in the N half of the sky...a cosmic snowball...resolution begins to be apparent in telescopes of 4-inch or larger." A diameter of up to 27' has been reported in instruments of 10-inches or more on exceptional nights - larger apertures filling in the faint outlying stars & increasing the apparent visual size beyond that seen in small glasses. Simply marvelous in 10- to 14-inch instruments! Anytime the author sees a big stellar beehive like M5 or M13 or M22, I can't help but think of the words of the poet Gerard Manley Hopkins: "Look at the stars! Look, look up at the stars! Oh look at all the fire-folk sitting in the air! The bright boroughs, the quivering citadels there!" It's this "aesthetic" approach to stargazing that provides its deepest meaning & greatest joy! "Remarkable GC...a treasure house...celestial splendor...one of the most compelling objects...dazzling mass of spark-ling stardust." The weak DS 5 SER (5.1, 10.1, 11") lies in the same field 20' to SE, providing a convenient focus check before looking at M5 itself - something that should always be done prior to viewing any GC or faint OC. This colossal radiant globe is "One of the most ancient clusters known" - some 13 *billion* years old! D=25,000LY

M16 SER/IC 4703 (13/0)	18 19 -13 47	OC DN	6.0 ---	25' 35'x28'	! **Eagle/Star Queen Nebula-Cluster** "A 'Jekyll & Hyde' object." M16 has usually been regarded by observers as just a "Grand cluster" of 60 or so stars. It's actually "A large scattered star cluster immersed in a vast diffuse nebula, a most wonderful object whose full glory is only revealed on long-exposure photographs." These show "A great complex of beautiful nebulosity threaded among the cluster stars" - partly in a striking "Outstretched 'eagle' shape." The NGC itself describes it as only a cluster, with no mention of nebulosity. "Until recently...the nebula...was considered a difficult visual challenge. But thanks to the widespread use of contrast-enhancing nebula filters (& of large-aperture telescopes among amateurs), the Eagle may now be seen to soar where it had never been seen before."! In 5- to 8-inch glasses, M16 appears as an "Irregular, hazy-looking star cluster...scattered over an area the size of the full Moon - embedded in the Eagle Nebula - too faint to be well seen in amateur scopes." "One of the most unusual objects in the sky & a fine sight at low X. Three nebulous regions in 4-inch." A "Hexagonal cluster of bright stars & star dust." "About 55 stars in a luminous field." "Scattered but fine, large stellar cluster. As the stars are disposed in numerous pairs among the evanescent points of more minute components, it forms a very pretty object." Seen "Through 11X80 binoculars paired with nebula filters, M16 is a bright smattering of stars engulfed in wonderful bright & dark clouds." The nebulosity itself is "Quite evident today in a good 6- or 8-inch telescope." "No more difficult than the nebula in the Pleiades."? One observer even claims that "Scopes larger than 8-inches show dark ('Bok') globules collapsing into protostars."!? As seen in its photos, an "Exceptional wonder of deep space. Thrusting boldly into the heart of the cloud rises a huge pinnacle like a cosmic mountain, the celestial throne of the *Star Queen* herself, wonderfully outlined in silhouette against the glowing fire-mist, where, as modern star pilgrims have learned, countless new stars are to be born. In the vast reaches of the Universe, modern telescopes reveal many vistas of unearthly beauty & wonder, but none, perhaps, which so perfectly evokes the very essence of celestial vastness & splendor, indefinable strangeness & mystery, the instinctive recognition of a vast cosmic drama being enacted, of a supreme masterwork of art being shown" as does M16! A faintly-fog-bound "nebulous star cluster" in an 8-inch at 60x! D=8,000LY
IC 4756 SER (2)	18 39 +05 27	OC	4.5	52'	Big splashy group of stars somewhat like nearby IC 4665 in OPH, needing a wide field for optimum view. "Unusually large naked-eye cluster, use lowest X." "Diameter 70'; scattered group of 80 stars." "70' extent! Very large & bright...great binocular object" - in which glasses this clan offers an "Impressive sight against the brilliant backdrop of the Autumn MW." A lovely spectacle in a 4-inch RFT at 16X! D=1,400LY
NGC 3115 SEX (12)	10 05 -07 43	EG	9.2	8'x3'	! **Spindle Galaxy** Well-named object! "An elongated bright nebula, extremes appear pointed." "Elongated outline & brighter center." "Very distinct with much brighter centre, bearing magnifying unusually well." A "Long narrow nebula...flashing stellar nucleus." "Glowing center." "Stellar nucleus clear at 285X (on 8-inch); outer flange visible." A "Bright...spindle with faint halo." An "Outstanding example of an EG -

superb object." "Very high surface brightness." Looks like a spiral to some observers: "Edge-on but no dust lane." "A faint needle of light against a starry backdrop...the galaxy's disk remains perfectly flawless, with no suggestion of any dust lanes or spiral shape" in 6- to 8-inch scopes. Yet "The best of modern photographs obtained with the 200-inch telescope appear to show a dual structure; there is a bright oval nuclear hub, & a thin flat equatorial plane...possibly a transition type E7-S0...no evidence for any true spiral pattern, nor any equatorial dust lane...so often seen in edge-on galaxies." "What a splendid sight NGC 3115 is" whatever its true nature! "The sole prize of Sextans." Easy in 3-inch glass - fascinating in big amateur scopes! D=21,000,000LY

Alpha TAU (8)	04 36 +16 31	SS	0.8-1.0 K5III	---	**! Aldebaran** The "Follower" or " Hindmost." Beautiful topaz gem projected against the Hyades Star Cluster (see below). "Red giant, appears to be part of the Hyades but is an unrelated foreground star superimposed on it by chance." Its lovely tint has been described as "brilliant orange", "red-orange", "gold", "rose", "rose red" & "pale red" by various observers. "Occultations of Aldebaran are not infrequent, as it lies in the Moon's way; they are striking phenomena, & to some observers are apt to exhibit the singular optical illusion of *projection*." "I have repeatedly seen it apparently projected on the disc of the Moon at the instant of immersion (for up to 3 seconds of time) when occulted by that body." While noted mainly for its brilliance & lovely hue, Aldebaran is also a weak DS. It has an 11th-mag. optical companion at 120" distant, described as orange, white or sky-blue & considered a light-test for a 3-inch glass. "B quite tough in primary's glare; requires clean, contrasty optics." (This little star is itself a tight 2" pair.) There's also closer in at 31" a very dim, 13th-mag. red dwarf CPM companion. D=65LY. Any discussion of Aldebaran must also include the following showpiece.
MEL 25/Theta-1/2 TAU (11/4)	04 29 +15 52	OC DS	0.5 3.4, 3.8 A7III, G7III	330' 337"	**! Hyades Cluster** Magnificent big, bright V-shaped naked-eye cluster that's a superb sight in binoculars & RFTs! "Two beautiful groups familiar to the first beginner in stellar astronomy, the Pleiades, & Hyades. Neither of these, however, is sufficiently concentrated to make a good telescopic object, excepting in an unusually large field." "Large bright cluster of about 200 stars covering 5 degrees of sky...brightest members form noticeable V-shape...best studied with binoculars rather than telescopes." "The 2nd closest OC (at D=150LY, after the UMA Cluster). A wonderful area to scan with binoculars." "Blazing V-shaped OC...a hundred blue-white class A suns, golden-solar & red Antarian giants." "Another old friend...there's much more to this cluster than meets the eye...of the 380 cluster members, about 130 are brighter than 9th-mag.... brilliant Aldebaran portraying the angry red eye" of the Bull & the V-shaped Hyades its horns! "Forms an attractive region for small telescopes; as a cluster it appears at its best in a good pair of binoculars." "An opera glass reveals many attractive pairs of stars." The most prominent of these - & the brightest cluster member - is the obvious naked-eye CPM duo Theta-1 & Theta-2 TAU in the bottom arm of the V. This "pretty pair" of white & yellowish giants is a lovely sight in both 10X50 binoculars & a 2-inch glass at 25X. Among the other doubles that will be seen in sweeping the cluster is the

wide (430") 5th-mag. blue-white combo Sigma-1 & Sigma-2 TAU. Located SE of Aldebaran, it appears on (3) showpiece lists. The entire Hyades complex is a joy to scan across in a 4-inch RFT at 16X! Among the best of the naked-eye star clusters.

| 118 TAU | 05 29 +25 09 | DS | 5.8, 6.6 | 5" |
| (8) | | | B8V, A1V | |

A nicely-paired, "well-matched double" for those who prefer their stars snug & close together! "Striking." "Beautiful." "Elegant." "White & rose, close, interesting pair." "Blue-white." "White, pale blue." Neat in 5-inch at 50X. CPM system.

| M1 TAU | 05 34 +22 01 | SR | 8.4 | 6'x4' |
| (20) | | | | |

! Crab Nebula If you're among the thousands of observers who over the years have used the author's *Sky & Telescope* reprint *The Finest Deep-Sky Objects*, I owe you (& the Crab) an apology! As the brightest supernova remnant in the sky, it very definitely should have been included as one of the 105 entries. (It *was* seriously considered but unfortunately never made it onto the final roster.) The "Famous...supernova remnant" from 1054AD colossal stellar explosion, with rapidly-spinning (33 times per second!) neutron star/pulsar at its core & still expanding at nearly 1,000 mps! "Celebrated... despite its fame, a disappointing object for small telescopes, appearing as an elliptical wisp of nebulosity." "Haze surrounds brighter middle...appears slightly greenish... streaks throughout inner portion extremely difficult" to glimpse in 4-inch. "Irregular shape, elongated SE-NW." "Oblong; pale; 1 degree NW (of) Zeta. Granular in 9.3-inch speculum (metal-mirrored reflector). First seen by Bevis, 1731. Its accidental re-discovery by Messier, while following a comet in 1758, led to the formation of the earliest catalogue of nebulae." "Famous 'Crab Nebula' of Lord Rosse (the analogy presupposed a strong Irish imagination), fairly defined tho pale, easily found...first object to induce Messier, in 1758, to compile his immortal catalog with a 2.5-inch comet sweeper!" "A somewhat egg-shaped body." It's "Dimly lurking between the horn-tips of Taurus." "Serrated outline visible in large instruments only." "Fringy appendages around & deep bifurcation to S." "Fine object, large, irregular & pearly white." "Sky's best example of a SN remnant." "A showpiece of the Winter sky." "One of the most fascinating & mysterious objects found anywhere in the heavens. Rapidly beating heart of an ancient star." "Small- to medium-aperture telescopes will show no more than...an 8th-mag. grayish smudge with little or no detail...intricate irregularities...begin to appear in 10-inch & larger instruments (which) increase the mottled look of M1, though the famous crablike appearance is difficult to detect visually." The granular appearance mentioned above caused John Herschel to call this object a "Barely resolvable cluster." Lord Rosse regarded the filaments (which he said resembled the legs of a crab in his giant reflector) as star chains & felt that in a bigger telescope this object "would then assume the ordinary form of a cluster." (Back then, it was believed that *all* nebulae were simply unresolved masses of stars!) "M1 is a fairly easy object, detectable in 3- & 4-inch apertures, appearing irregularly oval in a 6-inch glass, & showing some hint of detail in a 10-inch & larger instrument." It has been seen in binoculars & is easy in a 3-inch at 30X once you know what to look for. Adding to the interest of this object is the tiny, neat close DS Struve 742 (7.2, 7.8, 4").

Lying just 1/2 degree W of M1, it seems little-known to observers preoccupied with viewing the Crab itself. You'll definitely be delightfully surprised upon seeing it - "I didn't know that was there!" being the typical reaction. Again, that 3-inch shows it to a trained eye in good seeing at 30X & it's unmistakable at 45X. The Crab itself was often classified as a PN in the past, before its true nature was realized. It shines by the eerie bluish glow of "synchrotron radiation" & is presently about 7 LY in diameter. "Featureless gray ghost" in small scopes. "A telescopic treasure."! D=6,300LY

M45 TAU (16)	03 47 +24 07	OC	1.2	110'	

! **Pleiades/Seven Sisters Cluster** The most radiant, best-known & finest OC in the sky for viewing with the unaided eye, binoculars & RFTs! "Undoubtedly the most famous galactic star cluster in the heavens, known & regarded with reverence since remote antiquity. Commands admiration & attention. One of the most attractive celestial objects...the most dazzling view...is to be obtained in 20X70 binoculars, though...superb even in 7X50s. In a dark sky the 8 or 9 bright members glitter like an array of icy blue diamonds on black velvet; the frosty impression is increased by the nebulous haze which swirls about the stars & reflects their gleaming radiance like pale moonlight on a field of snow crystals." "The 6 principal stars of the Pleiades are evident to any clear sight; but glimpses of more are easily attainable (claims of up to 18 are on record!). A beautiful triangle of small stars will be found near the lucida, Alcyone. I have noticed the remarkable absence of colour in the group, except in one minute ruby star, & an orange outlier." That's simply because "None of its stars have evolved into red giants yet" due to its youth - only 20 million years old! "I'm always taken by the fact the dinosaurs never saw the Pleiades." "Conspicuous naked eye group ...glorious in 7x50 binoculars & in a 2-inch refractor at 15X one of the finest sights in the heavens. Less striking in larger scopes." "The loveliest of clusters." "Charming." "Celebrated group...beautiful. A miniature constellation." This clan looks like a tiny dipper, causing it to often be confused with the Little Dipper by novices! "Dazzling OC...spectacular. Through 11X80 glasses, the scene nearly left me breathless! Across the field were strewn dozens of stellar diamonds & sapphires." And many of these are in combination, for "Double & multiple stars are common in the Pleiades." One of the best involves Alcyone (Eta TAU) cited above, which is "Attractive for small instruments with a neat 1' triangle of 9th-mag. stars lying just 3' to the NW; Olcott calls this 'a quadruple star & a very beautiful object'." (This is including Alcyone itself, which has been called "The Light of the Pleiades.") "Celebrated cluster, brightest & most famous...in the sky. Under very clear conditions, the brightest part of embedded nebulosity, around Merope (Tau) may be glimpsed." This elusive 30'x30' "Tear-shaped glow in Pleiades" is known as the **Merope** or **Temple's Nebula**, NGC 1435. It shows as "A faint, extended, somewhat triangular haze, involving at its N extremity, Merope, the bright star SW of Alcyone." "A faint haziness involving some of the Pleiades." "Brilliance of the stars...makes seeing surrounding nebulosity difficult, but clear skies, a good WF eyepiece & (light-pollution) filter make the view staggering" in an 8-inch. This faint glow was described by its discoverer in his 4-inch as "Resembling a faint stain of fog, like the effect of 'a breath on a mirror'." "Only moderately difficult in good 6- & 8-inch instruments...surprisingly distinct...but totally invisible in moonlight." One highly experienced observer in exceptionally clear mountain air described this view through an 8-inch: "When I looked into the eyepiece, expecting to see a few faint wisps, the field was laced from edge to edge with bright wreaths of delicately structured nebulosity." This "Entire star-swarm is enveloped in a faint diffuse nebulosity of vast extent (Tennyson's "silver braid" from his immortal poem about the Pleiades) which appears to shine by reflected light." Note here, however, that dirt or moisture on the telescope optics - especially breath on the eyepiece itself - can cause *all* of the cluster stars to be surrounded with "nebulosity"! Since only six of the "Seven Sisters" are "bright with piercing sparkle" to most eyes, the legend of the "Lost Pleiad" has arisen - it invokes the theory that the strange variable 'shell star' Pleione was once much brighter than it now is. "Marvelous...splendid...delightful." "Most remarkable cluster in the heavens, the incomparable Pleiades." Like the neighboring Hyades, the "Classic Pleiades" cluster is strangely not listed in the NGC & carries the designation MEL 22. Membership in this "Swarm of fireflies" is variously estimated at between 250 & 500 suns. Gaze at it & "You too can behold this lovely vision."! D=410LY

(Please see Page 96)

Iota=6 TRI (14)	02 12 +30 18	DS	5.3, 6.9 G5III+F5V, F6V	4"	! A lovely, tight yellow & blue combo - split in small glass, but whose beauty increases steadily with aperture. "Golden yellow with close bluish companion." "Gold-yellow, bluish-green." "Sapphire & gold." "Topaz yellow, green; exquisite object." "Fine contrast." "Bright golden yellow pair." "Yellow/white." "Beautiful." "An attractive DS with a noticeable color contrast." Likened to Alpha HER, "but smaller & not so bright." Best seen in 6-inch or larger telescope. Both stars are spectroscopic binaries, so Iota is actually a quadruple system - each pair slowly orbiting the other. D=200LY
M33 TRI (19)	01 34 +30 39	SG	5.7	62'x39'	! **Pinwheel/Triangulum Galaxy** This tiny constellation's "One feature of superlative interest. Over 1/2 degree in extent, with crosses, rifts & nebulous condensations in a sea of glory!" "One of the best-known - & potentially most frustrating - galaxies in the sky." "Marvelously photogenic but hardly more than a grey cloud in small telescopes - you can go right over it without ever knowing you were there." "Among amateur observers it has been the source of the most discordant reports to be found in astronomical literature. While some find it easily visible in field glasses, or even with the naked eye, others report complete inability to locate the galaxy at all, & conclude that its must be incorrectly charted...the observer is looking for a much smaller & brighter object, rather than a dim glow comparable in apparent size to the Moon." It's the old 'low surface brightness' situation again! "Supposed to be a notoriously tricky object. Requires very dark skies - but what's everybody's problem? Plain as day in a 6-inch at 46X, filling nearly half the field with a lovely glow." "Very large, faint, ill-defined, visible from its great size. A very curious object, only fit for low powers, being actually imperceptible, from want of contrast, with my 144X" in a 3.7-inch glass. Huge - 15X on 2-inch & 30X on 3-inch show it plainly if you know what to look for! Fascinating

in 4-inch RFT at 16X, & a wondrous sight in 6-inch or larger instruments having at least a 1-degree apparent field of view! Here's an object that definitely 'grows on you' the more you see it! "Almost face-on, larger than the full Moon. Despite its size, & proximity, not prominent visually because of low surface brightness." "Large, diffuse spiral; requires dark sky." "Face-on spiral." "7x50 (binoculars) show it rather easily but very faint & difficult in 4-inch. Three-D effect sometimes seen in field glasses & short-focus instruments when the air is not steady." "Can look lovely in binoculars & RFTs on a dark night. If the Moon is up, don't even waste your time trying!" "While it's beauty is undeniable, has also earned the reputation of being notoriously difficult to find (as well as view!). Most observers see M33 as little more than an ill-defined blob at first. As the...eye becomes more accustomed to the galaxy's appearance...it begins to yield riches unsurpassed by any other galaxy N of the celestial equator...two delicate spiral arms can be seen unwinding from the core (& also) numerous bright pockets of nebulosity" in 8-inch & larger scopes. The brightest of these is NGC 604, located 10' NE of M33's nucleus - one of the many starclouds/H-II regions in its spiral arms. "I find it ironic that NGC 604 is one of the brightest nebulae in the midautumn sky, yet it is not even in our galaxy...stands out surprisingly well. In fact, under slightly hazy skies, the nebula may actually be visible while the galaxy is nowhere to be found!" It shows as an 11th-mag. "soft blur" about 1' in size. M33 itself displays "Huge ghostly spiral arms...resembling a blanket tucked around the galaxy's central hub" while many "Irregular nodosities give it a curdled appearance." "Large, distinct, but faint & ill-defined, pale white nebula" is how most observers find M33 on their first encounter. But as the foregoing descriptions indicate, there's much more to be seen here than first greets the eye! "M33 & M31 are only about 500,000LY apart (in space). Stargazers in the Andromeda Galaxy should have a lovely view of the Triangulum Galaxy, & vice versa."! M33 shares the name "Pinwheel" with both M 99 in COM & M101 in UMA. "Marvelous M33." Nearest of all the spirals after that in Andromeda: D=3,600,000LY

Zeta/80 UMA (20)	13 24 +54 56	DS	2.3, 4.0, 4.0 A2V, A1, A5V	14", 709"

! Mizar & Alcor Finest "white" DS in the heavens, with Alcor 12' distant. All three suns look like radiant blue-white diamonds - superb sight even in 2-inch glass at 25X! Famous as "the DS at the bend in the Big Dipper's handle." Both "Greenish white... fine pair...forms a noble group with Alcor...a pair to the naked eye, & thus becomes an excellent object for a beginner, as the telescopic increase of brightness & distance (separation) admits of direct comparison." Thus, at public star parties, this threesome has become *the* standard object for comparing what the unaided eye sees with what a telescope reveals. "The most famous naked-eye DS in the entire sky. In reality, however, the team of Alcor & Mizar is illusory since they are actually nowhere near each other in the sky." (Latest values D=92LY & 59LY, respectively; distance given for all three stars until recently was 88LY?) Even if "Not an orbiting member", Alcor is drifting through space with Mizar, forming a very wide CPM pair. The latter duo is an ultra-long-period binary whose orbit requires several thousand years to complete. "Striking pair of beacons through all telescopes." "Brilliant white & pale emerald." "Both white." "For over three centuries it has remained one of the most celebrated DSs, & is often the first example of that class to be observed by many a modern amateur...an easy & striking pair." "Celebrated multiple star"; the first discovered telescopically (1650), the first to be photographed (1857) & the first spectroscopic binary ever found (1889)!! (Actually, Mizar A & B, & Alcor are *all* spectroscopic pairs!) "Pioneer star. The pair, so happily placed in the crook of the Big Dipper's handle, never fail to inspire awe, however frequently observed." "Splendid...truly fine object...requires but little optical aid to divorce the components...must not be quitted without a notice of Alcor." This historic pair is "Just about the easiest DS to find", "Probably the best-known DS in the heavens & one of the most beautiful" & also "One of the most observed...doubles in the N sky."! Magnificent across the entire 2- to 14-inch aperture range of survey!

Xi UMA (15)	11 18 +31 32	DS	4.3, 4.8 G0V, G0V	1.8"	

! Another famous pair! The first binary to have its orbital period (60 years) computed, in 1828. Has made more than 3 circuits since it discovery in 1780! Last at maximum separation of 3" in 1975, it closed up to minimum at 0.8" in 1992 & is now opening again. These nearly identical suns are currently tight in 3- & 4-inch glasses. "Two yellow components...at their closest require apertures of 6-inches to split them." Xi was "The subject of the first experiment in the extension of Newtonian principles to the sidereal universe." "Fine close binary of great historical interest. No very definite color contrast exists in this pair; to most observers both stars appear a clear pale yellow." "Both white." "Extraordinary pair...subdued white, greyish white, & both very bright." "Almost touching." "Close but easy" in 8-inch. Neat in 5-inch at 100X. Both stars are spectroscopic binaries, so Xi is really a double-double system! D=26LY

M81/M82 UMA (20/20)	09 56 +69 04	SG IG	6.9 8.4	26'x14' 11'x5'	

! **Bode's Nebulae** The finest galaxy-pair in the sky, easily seen even in the smallest of telescopes! "Two nebulae 1/2 degree apart; M81 bright, with vivid nucleus, finely grouped with small stars, two of which are projected upon the haze. M82 (Bode's Nebula), curious narrow curved ray...two nuclei, & sparkling as if resolvable, which M81 is not." "Argent, 30' apart. Unlike in size, structure & glow. Worth the search." "Stunning pair of galaxies." "A circular nebulosity. (&) Has the effect of a narrow curved streak, somewhat like a scimitar blade." "A fine bright oval nebula, of a white colour. (&) Very long, narrow & bright - rather paler than no. 81." "Bright & huge oblong; faint spiral arms; bright stellar nucleus; traces of dark lanes visible at high X. (&) Bright & large slash...much detail visible; splotchy & dotted" in 8-inch. "Bode's Nebulae. Bright with fine elliptical form, having distinct nuclear area...photos picture it with extraordinary beauty, outer whirls showing strong spiral movement. (&) Long, narrow & fainter, but an exceptional object, crossed by dark bands." (The name Bode's Nebula actually refers to M82, but is often used to include both galaxies - & sometimes interchangeably: "Bode's 'Nebula' - beautiful, paired with M82.") "Form a pretty pair in low X field, 38' apart." "Duo of magnificently dissimilar galaxies." "Astronomers living in one of these galaxies will have a fine view of the other one!" Truly "One of the most intriguing pairs of galaxies found anywhere in the sky. Classic example of an

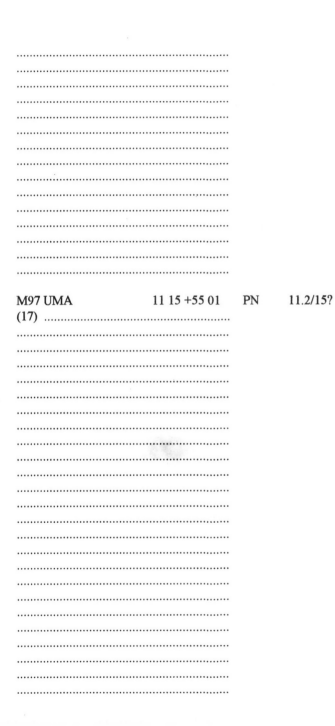

Sb spiral. (&) Strikingly different kind of beast - the brightest & most famous of its species." "Classic spiral. (&) 'Exploding' galaxy...an intriguing edge-on irregular." M81 is a "Beautiful roundish, softly glowing patch, one of the brightest in the sky." "Splendid spiral - a beautiful object! Has most strongly granular central region of almost any galaxy" as seen in 4-inch. "Outstanding...spiral. Once spied, M81 will put on a memorable performance. Though it is about a full magnitude fainter than its neighbor, M82 offers more structural detail...most prominent feature...is a jagged rift of darkness that rips across its center." "An extended ray." "A gem! In low X field forms a beautiful pair with M81. A silver sliver. Dark bands not seen" in a 4-inch. "Very bright edge-on with dark cuts." "Bright spindle-shaped galaxy of unusual type, forming a striking pair with M81." "Appear like an elongated blur...both...related." This "dynamic duo" is quite easy in a 3-inch at 30X, fascinating in a 6-inch at 50X & truly stunning in a 9-inch at 70X! M81's nuclear hub & spiral-arm halo are in lovely contrast with cigar-shaped M82 & its dust lanes, as these two vast star cities serenely sail the ocean of intergalactic space together! "Pair of show objects." D=7,000,000LY

M97 UMA 11 15 +55 01 PN 11.2/15? 180"
(17)

! Owl Nebula At an Astronomical League meeting over 35 years ago, the author gave his first paper on the survey that ultimately led to this compilation. I began by stating that "The Owl Nebula is for the birds!", due to early difficulties in seeing it. But since then, I've come to really like this nocturnal feathery bird with the big eyes (whose real-life counterpart's view of the heavens must be fantastic!). "Harder to see than this (its listed mag.) might suggest. Rather large grey oval, practically featureless, tho there is a slight indication of two dark areas" (the 'eyes') in a 4-inch. "Large but dim PN." A "Large pale PN; very remarkable object...Earl of Rosse (saw) two large perforations (whence it has been called the 'Owl Nebula'). Before 1850 there was a star in each opening; since, one only."! Curiously, John Herschel doesn't show the 'eyes' in a drawing made with his 18-inch reflector & described M97 as "Light nearly equable, though with softened edge, & faintly bicentral." And his father, William, saw it as "A body of equal light throughout." "Nesting under the Big Dipper." "Large, round planetary...looking somewhat owl-like." "Round, with two low-contrast 'holes'." "One of the most challenging PNs for small scopes...one of only four in the Messier catalog & one of the faintest M objects...'eyes' visible in 10-inch with averted vision...resembling eyes staring from the inky blackness of space." "A 6-inch will show something of them." "Dull-toned but memorable." "Bright & round; N-filter shows...detail including two asymmetric inner voids" in 8-inch. "The central star, about 14th-mag. visually, lies on the central 'bridge' of luminosity which separates the two dark vacancies. To most visual observers today it is merely a large pale disk of light. With great telescopes, there is a distinct impression of a spectral bluish-green tinge." "A large PN or globular collection of nebulous matter...very singular object...circular & uniform, & after a long inspection looks like a condensed mass of attenuated light, seemingly of the size of Jupiter (?)." "Double formation. Gaseous. Five faint central stars (in a 10-inch). Region abounds in nebulae." Closest of these to M97 is M108, a 10th-mag.

edge-on spiral measuring 8'x2' in size. "Renowned. The Owl is perched just 48' SE of M108, possibly within the same low-X field...8-inch telescope begins to add a little personality...by showing hints of two dark ovoids either side of center." M108 itself "Resembles a nebulous cigar nestled among an attractive field of stars." "Seen nearly edgewise - a silver-white beauty for small telescopes, saucer-shaped & fairly well defined." "Large, mottled edge-on with foreground stars." Appears on (6) showpiece lists. This "Low-power pair" & "Celestial odd couple" easily fits within the field of a 6-inch RFT at 24X. M97 resides well within our Galaxy at a distance reckoned in thousands of LYs (the exact figure is highly uncertain, values ranging from 1,300LY to 10,000 LY having been published!), while 'neighboring' M108 lies far beyond the Owl & confines of our MW at an intergalactic distance measured in *millions* of LYs!

M101 UMA (9)	14 03 +54 21	SG	7.7	27'x26'	

! Pinwheel Galaxy Huge, rather dim spiral reminiscent of M33 (TRI), with whom it shares its descriptive name. "Fantastic...spiral." "Owners of small telescopes must be content with glimpsing it as a 9th-mag. smudge, more than half the Moon's apparent diameter." "Very large bright spiral seen face-on. Beautiful object. Clearly seen central region has a fluffy texture & silvery hue. Surrounding this area is a soft sheen containing some nebulous patches." "Nice face-on, spiral structure & knots in dark skies." "A large & faint nebula." "Huge & bright! Get ready to spend the night!" "Pale white nebula - under a favorable view it is large & well spread, though some-what faint except toward the centre, where it brightens. One of those globular nebulae that seem to be caused by a vast agglomeration of stars rather than by a mass of diffused, luminous matter...the paleness tells of its inconceivable distance." Written in 1844, this comment by Smyth truly "Reveals a remarkable prophetic insight" as it anticipated the real nature of galaxies nearly a century before it became known! "Large open spiral - weak spiral shape in elusive halo of nebulosity." "A blur in small scopes. Bright knots in an amorphous field with larger ones. Giant Sc spiral." "Although its total brightness is great, its light is so spread out that it is difficult to see in small instruments." "Due to low surface brightness, many observers scan right over this... spiral without even realizing it." But "Even though its low surface brightness makes it infamously difficult to locate...M101 holds a wealth of detail for patient visitors. Indeed, patience is a prerequisite to finding this huge face-on spiral galaxy in the first place. When initially encountered, (it) will probably strike you as little more than an amorphous featureless glow...on closer inspection this initial blandness blossoms into great intricacy. Under exceptional skies, an 8-inch telescope reveals a subtle spiral pattern within the exceedingly faint halo." As with its counterpart M33, many star-clouds/HII regions can be glimpsed as clumps of brightness within this spiral arm halo, more than a dozen having their own NGC numbers. There's also at least one relatively bright star within M101's glow "that might cause your heart rate to jump thinking you've found a supernova...(but it's) only an interloper."! A number of such outbursts *have* been seen in this system & it's a good target to monitor for future ones. Included by Smyth in his *Cycle of Celestial Objects* (*The Bedford Catalogue*) under

....				
....				
....				
....				
....				

BOO, M101 was strangely overlooked by Webb in his classic *Celestial Objects for Common Telescopes*! Visible in a 3-inch glass at 30X on a dark night, the best views come in 10-inch or larger apertures using eyepieces giving nearly a full degree field of view & magnifications around 60X. "Splendid large face-on spiral."! Most Messier authorities believe that M102 is a duplicate observation of M101. D=15,000,000LY

NGC 2841 UMA	09 22 +50 58	SG	9.3	8'x4'
(6)				
....				
....				
....				
....				
....				
....			(Please see Page 96)	
....				

Fine specimen of the host of other, non-Messier, galaxies to be found scattered across this constellation. "Oval, lambent, with clear nucleus." "Classic elongated spiral; very bright." "Large, bright spiral, look for dusty structure." "Fine spiral with symmetrical whorls." "Not as popular among observers as some of the other UMA galaxies...nevertheless bright enough to be seen in 3- & 4-inch telescopes...a 9th-mag., cigar-shaped glow accentuated by a bright, oval core." "Large, oval, bright, with nucleus. Dull object, (in) 3.7-inch, in fine field." "Impressive galaxy in small scopes; 3 supernovae." "Bright nebula of a pale creamy whiteness. Large, nucleated & elliptical" shaped. "A classic...spiral." Fine sight in 8-inch at 60X - as good as many of Messier's galaxies!

Alpha UMI	02 32 +89 16	DS	1.9-2.1, 9.0	18"
(15)			F7I	
....				
....				
....				
....				
....				
....				
....				
....				
....				
....				
....				
....				
....				
....				
....				
....				
....				

! **Polaris/Pole/North Star** One of the best-known & most useful stars in the sky - that marking the position of the North Celestial Pole. Also one of the select few deep-sky objects that 'does something' in the eyepiece - in this case, a 'binary' with a *24-hour* 'orbital period' that you can watch turning before your eyes! This motion is, of course, only apparent, being caused by the spinning Earth rather than by the DS itself. A good plan is to begin your evening observing session with a quick glance at the direction of the companion on the sky (its 'position angle', with N being 0 degrees, E 90 degrees, etc.). A few hours later, at the end of your viewing session, take another look - you'll find that the secondary has noticeably moved around the primary! "Yellow supergiant Cepheid...companion visible in small telescopes." The small star can be glimpsed in a 2-inch but a 3-inch is needed to make it definite under typical conditions. "Difficult faint companion." "Not an easy object." A "Common test, but only suited for small apertures, being easy with anything much exceeding 2-inches." "Topaz yellow, pale white." "Yellowish-white & blue." "White/blue-white." Polaris makes a convenient star for testing optics, being not too bright or faint & it doesn't move out of the field if there's no drive! "The most valuable star in the heavens." Currently about 45' from the true pole, Polaris will approach to within 27' in the year 2100 due to precession. Most know that Vega was the Pole Star 12,000 years ago & will again assume that role 12,000 years hence. But you may be unaware that some 425,000 years ago Earth had a brilliant *double Pole Star* when Capella & Aldebaran, only a degree apart, occupied that position! Polaris is a CPM pair & the primary a spectroscopic binary, as well as a Cepheid variable star - in fact, the brightest one in the entire sky! D=820LY

Gamma VEL	08 10 -47 20	DS	1.8, 4.3	41"
(5)			WC8+O8, B1IV	
....				

! Although more than 2 degrees below our survey's limit, this stunning object is being included for the benefit of observers living in the S or traveling there (as perhaps to the annual Winter Star Party in the Florida Keys!), because it is indeed "One of the most beautiful objects in the heavens."! "Blazing pair in MW field." "Beautiful easy pair." "A wonderful five-star system for backyard telescopes." "Interesting multiple star... two blue-white companions...also two wider (ones)." This includes 8th- & 9th-mag. A-type stars at 62" & 94", respectively; all of these look much fainter than the values given due to heavy atmospheric extinction so near the horizon. But the AB pair comes shining through anyway! "A splendid & easy DS for any small telescope, resolvable in...binoculars." The colors of the three brightest components are given as "White, greenish-white & purple." "Very beautiful." For those using eyepiece spectroscopes, the spectrum of the primary shows "Extraordinary beauty. An intensely bright line in the blue, & the gorgeous group of three bright lines in the yellow & orange, render the spectrum...incomparably the most brilliant & striking in the whole heavens." Gamma is located "In a fine starry field in the MW" & is an amazing sight in small to medium apertures at low magnifications. Fixed pair. Primary also a close spectroscopic binary (note dual spectral type). Distance uncertain: D=500LY to 1,600LY variously given!

NGC 3132 VEL (8)	10 08 -40 26	PN	8.2/9.5	84"x52"	**! Eight-Burst Planetary** One of the brightest & best planetaries in the sky, yet rarely looked at due to its low DEC. "Located squarely on the Vela-Antlia border. More or less comparable in size to the Ring Nebula...appears more conspicuous than the Ring, owing to the presence of the unusually bright central star, listed as either 9th- or 10th-mag. The disk is noticeably elliptical...with much diaphanous detail & a structure suggesting the appearance of several oval rings superimposed & tilted at different angles...'bright white' & without a sign of the usual blue-green tint common in the planetaries." "Easy 9th-mag. central star." "Bright disk of nebulosity surrounding a 10th-mag. star." "Bright, oblong; requires excellent S horizon." "Must rank as one of the top 10 planetaries." "Unusual multiring structure...best seen in...larger backyard instruments...slight bluish tint." Cloud & central sun easily visible in 4-inch at 45X, & an astounding sight in 10-inch or larger scopes at 100X & over! Nebula white with a hint of blue. The "Southern Ring Nebula." Don't miss this beauty! D=2,000LY
Alpha VIR (2)	13 25 -11 10	SS	0.97 B1V+B2V	---	**Spica** "Ear of Wheat" Star. Icy-blue-looking member of the 'First-Magnitude Club' - lovely tinted jewel even to the unaided eye! A "Blue gem." "Extremely beautiful pure white star." "Brilliant flushed white...beautiful bright star...in a clear dark field, &... insulated." A "Brilliant 'helium' type star" some 2,300 x our Sun's brightness! Spica is a visually unresolved spectroscopic binary with 4-day period, which is "Remarkably short for stars of the giant class." The two suns also slightly eclipse each other during their orbital dance, causing a visually-unnoticeable drop in brightness. D=280LY
Gamma VIR (21)	12 42 -01 27	DS	3.5, 3.5 F0V, F0V	1.8"	**! Porrima** Famous bright binary system with 171-year period that's now becoming tighter with each passing year! "This splendid binary is notable for the brilliance of its components...& for their conspicuous orbital motion. When this pair is widest (6" - last in 1919), it's easy even in very small amateur telescopes, but when closest (next in 2008 at 0.4") it appears single in all except the largest instruments." Near 'periastron'

(closest approach of the stars), the position angle changes by more than *70 degrees per year*! The blended image of Porrima appears then as a rotating egg in 8-inch & larger scopes at high X. "Celebrated DS...matching pair of yellow-white stars." "Beautiful look-alike pair." "This wonderful pair has been widening ever since they closed up out of all telescopic reach in 1836" (written in 1859). "I was astonished, on gazing at its morning apparition in January (1836), to find it a single star! In fact, whether the real discs were over each other or not, my whole powers, patiently worked from 240X to 1200X (on a 6-inch), could only make the object round." But "Within a few months the pair had widened & soon became an easy object for small telescopes."! "Superb binary." "Splendid." "One of the finest pairs visible!" "A beautiful object." "One of the classical DSs." The two suns appear "Clear pale yellow, & look for all the world like the remote twin head-lamps of some celestial auto, approaching from deep space." "Well worth taking a break from the faint fuzzies (Virgo's galaxies) to see." "White & yellowish, equal brightness, noticeable color difference." "Both yellow." "Silvery white, pale yellow. So wonderful an elliptic motion." Currently very tight in 3- & 4-inch glasses, but still relatively easy in a 5-inch under good seeing. Actual distance between the two stars varies from 400 million to 6.3 billion miles! D=32LY

Theta VIR (10)	13 10 -05 32	DS	4.4, 9.4, 10.4 A1V	7", 70"	Delicate magnitude-contrast trio with subtle hues for mid-size scopes. "White, violet, dusky. Fine triple, but difficult with 3-inch." "Blue-white components visible in small scopes." "Beautiful pair" in 8-inch. The A-B duo's tint's have been given as "yellow & red", "white & orange", "both white" & "white & greenish or bluish." CPM group.
SS VIR (2)	12 25 +00 48	SS	6.0-9.6 N (C5)	---	Ruddy carbon, Mira variable with 355-day period that's "Notable for its deep red tint." "Brilliant red-orange." When bright, color obvious even in 3-inch - striking in 6-inch!
M84/M86/M87 VIR (9/8/7)	12 25 +12 53 12 26 +12 57 12 31 +12 24	EG EG EG	9.3 9.2 8.6	5'x4' 7'x6' 7'x7'	

! Coma-Virgo Galaxy Cluster Truly "A wonderland surpassing the dreams of fiction writers is hidden in the Coma-Virgo region of the sky, between the stars Vindemiatrix (Epsilon VIR) & Denebola (Beta LEO)." This area is "Especially remarkable for those possessed of adequate optical means, on account of the wonderful *nebulous region*, in which a far greater number of these extraordinary bodies are accumulated than in any other equal area of the heavens. Few of them, however, are individually interesting; it is the mysterious thronging together of these objects, whatever may be their nature, that opens such a field for curiosity. They are in general so much alike that I have only adduced a few as specimens." "A vast Sargasso sea of star-illuminated cosmic matter." "In wondrous nebulous region...marvelous as sweep...identification difficult." "A galaxy hunter's paradise." "Welcome to galaxy country! 2,500 individual galaxies... spanning over 100 square degrees" - of which several hundred can be glimpsed with backyard scopes & more than a dozen with binoculars! "Extraordinary conglomerate... a vast laboratory...broad grand stratum of nebulae, this glorious but most mysterious zone of diffused spots. Several of the finest objects of Messier & the Herschels will readily be picked up by the keen observer in extraordinary proximity." "One of the most remarkable areas of the heavens. A celestial wonderland of innumerable star cities. In small telescopes, none of the members of this galaxy cluster are visually impressive; they appear mainly as pale patches of light, round, elongated, & irregular; even the brightest members are not brilliant or striking objects.(?) It is the knowledge of the actual nature of these glowing spots that compels the...amateur to return again & again...to observe & contemplate a celestial panorama surpassing the highest flights of human imagination. Here in the Virgo Cloud one may gaze upon the radiance of a hundred vastly remote star cities, twinkling across the millions of lightyears...at least a 6-inch telescope is recommended; a clear & moonless night is essential...30X to 50X is ideal on 6- to 8-inch instruments." "Holds a lifetime of fascination." While very slow sweeping is the best way to see & enjoy this cluster, a sampling of its brighter members will convey something of its amazing treasures. The three Messier galaxies listed here lie at the core of this "Confusion of silver sands" & some of its other outstanding units are described below. M84: "A beautiful sight, easy target. More conspicuous than M86 in same field." "Very bright elliptical with bright core." "In richest part of C-V galaxy cluster." "M86 & many NGCs nearby; lots to explore!" M86: "Giant elliptical. Impressive in small telescopes." "20' E of M84, more diffuse core." M84 & M86: "Both appear in the same telescopic field as fuzzy patches with noticeably brighter cores." "Pair of almost identical elliptical galaxies." "NGC 4388 lies 16' S & NGC 4402 10' N." "Nine galaxies in 1 degree field centered on M84 & M86 comprise 'The Nonet'." M87: "The one with the famous jet & black hole." "Virgo's grand giant elliptical - a circular fuzzball." "Celebrated giant EG - a rounded 9th-mag. glow with noticeable nucleus." "Brightest member of the Virgo Cluster & intrinsically one of the brightest known." "Brightest elliptical in the Virgo Cluster, NGC 4478 10' SW." "Brilliant not-quite-stellar nucleus...engulfed in a fainter round mist...backyard 'scopes just don't have enough oomph to show the jet." The jet extends to the NW - it's been seen visually with the 100-inch reflector at Mt. Wilson & described as "bright" in the 200-inch reflector at Palomar! This colossal star city contains at least a *thousand* GCs! All three M-objects visible together in same wide field in 3-inch glass at 30X. "Downtown Virgo"! Distance to center of this vast "Realm of the Nebulae" = 70,000,000LY

M49 VIR (11)	12 30 +08 00	EG	8.4	9'x7'	! "Very bright elliptical." "One of the brightest members of the Virgo galaxy cluster." "Bright, round, & well-defined nebula. Has a very pearly aspect." "Notable nebula." "Faint haze in beautiful position between two 6th-mag. stars." "Bright with prominent core." "*Illustris, grandis.*" "Resemble a GC or the head of a comet. Bright central region more sharply defined than usual among galaxies." "One of the largest & most massive elliptical systems known." "Large, bright, round, & pearly, situated between two bright telescopic stars." Obvious even in 2-inch glass despite D=65,000,000LY!
M59/M60 VIR (6/9)	12 42 +11 39	EG	9.8	5'x3'	Nice galaxy pair, 25' apart, with NGC 4647 (11th-mag., 3'x2') also in field 2.5' NW! M59: "Bright elliptical paired with M60." "Miniature of M32 (AND)." "Bright oval with sharp nucleus, pair with M60 25' E." "Bright little nebula." M60: A "Double
		EG	8.8	7'x6'	

M61 VIR (7)	12 22 +04 28	SG	9.7	6'x6'
M104 VIR (15)	12 40 -11 37	SG	8.3	9'x4'

nebula." "Bright elliptical with M59 & NGC 4647", the latter a spiral appearing as "a fuzzy star." M59 & M60: Together forming a "Mysterious & shadowy doublet."!

All of the above M galaxies have been ellipticals, but there are also many spirals in the Coma-Virgo Cluster - this one being a perfect example! "Bright face-on with evident spiral structure." "Face-on, two-armed spiral." "One of the finest small spirals in the sky." "Large pale-white nebula, well defined...but feeble - blazes towards the middle." "A fine object for 4-inch", but like most galaxies best seen in 6-inch or larger aperture.

! Sombrero Galaxy One of finest edge-on spiral galaxies in the sky - so bright that it's visible even in Moonlight or moderate light pollution with 3- & 4-inch glasses! A hazy ellipse with dark equatorial band (giving rise to its name) in a 6-inch. "Look for dust lane." This 'hat brim' appears displaced somewhat S of the galaxy's big bulbous center. "Easy...bright...& large, tilted only 6 degrees (to our line of sight). Beautiful object in 4-inch for well-trained eye. Until the observer has gained experience...may look...featureless...'hat brim' difficult to study visually - appears to have a curdled texture." "Popularly known as the Sombrero Hat...the dark lane of dust around its rim needs apertures above 6-inches to be detected visually" - all depending on how trained your eye is & the sky conditions. "Finest galaxy in VIR. Dark lane visible in 3-inch. On a good night is quite a sight. Splendid edge-on (having) appearance of a spaceship from B-grade Sci-Fi thriller." "Bulging edge-on split by dust lane." "Bright spindle with dark lane." "Prominent equatorial dust lane" as seen in 8-inch. "Easy object for even the smallest telescopes (&) well within range of binoculars. Magnificent." "A lucid white elliptical nebula...in an elegant field of small stars. On intense attention, may be seen to blaze in the middle...but it seems a mere wisp of subdued light" (?) in 6-inch. "Prominent central core, broad spiral-arm rim, & conspicuous dark lane. The overall visual effect is that of a Mexican hat...viewed through a 4- or 6-inch telescope, the Sombrero's dark lane may be seen cutting across the S side of the nucleus. The added light-gathering ability of an 8-inch instrument extends the lane fully across the galactic 'brim' of M104." "The dark band is not exactly easy in a 10-inch aperture."!? (It's visible to the author with averted vision in a 5-inch at 80X under suburban skies.) "Long...nucleus & dark cleft. Beautiful low-powered field: fine & singular 7th-mag. group NW." "M104 lies in a very attractive field, centered in a sparkling group of six 7th-mag. stars; several of these, about 19' out from the galaxy to the WNW, form the multiple star Struve 1664. This is a 3.5' chain of three stars with the westernmost member a 26" pair (of 8th- & 9th-mag.)." This striking asterism has been dubbed "Little Sagitta" from its arrow shape - & it conveniently points to M104 some 20' E! It, in turn, lies just NE of another asterism just over the border in CRV known as the "Stargate" - a "Unique triangle-within-a-triangle pattern of six stars...a one-of-a-kind celestial sight." You're sure to run into both while sweeping for M104 itself! Located some 25 degree S of the center of the Coma-Virgo Cluster, this spiral is considered an outlying member - orbiting on the near side of the swarm to us. D=40,000,000LY

NGC 4762 VIR	12 53 +11 14	SG	10.2	9'x2'
(4)				

The Kite Neat example of the many Herschel/NGC galaxies to be found in the Coma-Virgo Cluster. "Beautiful, thin edge-on with intense core." "Flattest galaxy known; 4754 in same field." Both objects "In one field (with) 4762 like a paper kite; beautifully grouped with 3 stars." "Long bright ray, nucleus." "Edge-on; thick streak with extending tufts at end; 11' pair with 4754 (12th-mag., 2'x1')." "Pale, elliptical nebula ...from its superior brightness at the S, or upper, end it rises while gazing from the dumpy egg-shape to that of a paper kite: over it is an arch formed by three telescopic stars, the symmetry of which is so peculiar as to add to that appearance." Some sources list NGC 4762 as faint as 11th-mag. Its tiny companion is a barred spiral. Many other galaxy pairs & triplets will be found in sweeping this amazing realm of "faint fuzzies"!

3C273 VIR	12 29 +02 03	QUASAR	12.8	---
(1)				

* First Quasar The brightest, closest & first of its enigmatic class to be discovered. D=3 *Billion* LY, or nearly 18,000,000,000,000,000,000,000 miles from us! And yet, it's been glimpsed in telescopes smaller than 6-inches in aperture by many experienced observers & is definite in 10-inch & larger instruments! "Appears as a 13th-mag. blue star...brightest quasar." Its brightness varies rapidly & unpredictably by about a full magnitude. "One of the most brilliant single objects in the universe, & possibly the most distant object visible with a 10-inch telescope. It is located 3.5 degrees NE from Eta VIR, & has the appearance of a faint bluish 'star'." This amazing object is racing away from us in the overall expansion of the universe at a velocity (or 'red shift') of some *30,000 miles per second* - or 1/6th the speed of light! "It is within the grasp of a 6-inch telescope, but only on nights when the air is dry & the sky is crystal clear. Its blue color should help set it apart from stellar neighbors." Even when using extremely accurate digital setting circles, a detailed finder chart is required for positive identification. The author wrote the very first article ever published on *actually looking* at quasars, in the December, 1979, issue of *Astronomy* magazine; it contains such a map along with a high-resolution photograph of the field. (Others have appeared from time to time in the observing sections of both *Astronomy* & *Sky & Telescope*.). Here as in other areas of deep-sky observing, it's important to see with our minds as well as sight - it's what that faint glimmer of light in the eyepiece field represents that makes it so fascinating to star lovers. Consider that 3C273 has the combined energy output of more than a *thousand* typical galaxies, all radiating from a source only mere light-*days* in size! Also - "Imagine: the light we see coming from 3C273 tonight left there when single-cell life forms ruled the Earth!" And yet we can glimpse this tiny "blue dot of light" with backyard scopes! Quoting from the conclusion of my article (which lists several other quasars within reach of larger instruments): "When their baffling nature & possible grand significance in the cosmic scheme of things is appreciated, catching sight of a quasar can be a thrilling experience. It is nothing less than amazing that a small disk of glass can capture & bring into focus the ancient photons of light which have traversed the expanse of the observable universe on a journey lasting billions of years. Truly, the telescope is a magic 'time machine' which makes possible vast journeys across the cosmos in sight & mind at the turn of a knob!"

COL 399 VUL (2)	19 25 +20 11	AS	3.6	60'

* Coat Hanger Asterism/Brocchi's Cluster With the possible exception of the Big Dipper itself, the best & one of the most striking examples of an asterism in the sky! Dimly visible to the unaided eye - but too spread out for typical telescopic fields - it's at its sparkling best in binoculars & RFTs. "A notable little group of 6th- & 7th-mag. stars known colloquially as the Coathanger, a striking scene in binoculars...(its) most remarkable feature is an almost straight line of six stars; a curve of stars forming a hook extends from the centre of this line." "Situated in a remarkable little asterism, as seen with an opera glass; six stars in a line." "Scattered naked-eye group including 4, 5, 7 VUL" - its 3 brightest members. "Look for a smudge of grayish cosmic cotton speckled with a few points of light" to the eye. "With 7X binoculars, six stars aligned in a row appear to form the Coathanger's crossbar, while four others curve away to mark the 'hook.' In all, Collinder 399 contains some five dozen suns spanning a full degree of sky. Surrounded by a magnificent field filled with stardust, this is sure to become one of your favorite summertime objects." The Coathanger will very definitely 'hook' you on observing asterisms! D=420LY The OC NGC 6802 lies "At E end of Brocchi's cluster." Described as "very rich" & a "neat silver sprinkle", it's only 3' in size & shines at 9th-mag. - in marked contrast to the huge 'celestial coathanger' itself!

NGC 6940 VUL (7)	20 35 +28 18	OC	6.3	31'

"A splendid OC. All five dozen stars within this rich collection shine like sapphires against a black velvet backdrop, save for one renegade. That maverick, FG VUL, also happens to be the group's brightest star, a ruby red semiregular variable...& (it) offers a striking color contrast to its blue-white brethren." "100-plus 9th-mag. & dimmer members...large & medium rich." A "Rich star field, includes orange variable star." "Fairly rich cluster in MW." "A striking OC." FG itself varies from mag. 9.0 to 9.5 over a period of 80 days. An "Enchantress...beautiful contrast of colors." D=2,500LY

M27 VUL (21)	20 00 +22 43	PN	7.6/13.9	8'x5'

! Dumbbell Nebula Saving one of the most magnificent celestial wonders for last, this big, bright planetary is tied with the Ring Nebula (M57 in LYR) as perhaps the finest example of its class! "One of the showpiece sights of the sky, visible in binoculars on a clear night, large & bright, reputed to be the most conspicuous of its kind, 1/4th the diameter of the full Moon. Visually appears as a dumbbell-shaped, misty green glow." "Superb object." "Of all the PNs in our sky, none are as easy to see as M27. And the view...through a large telescope is sure to leave you breathless! Different observers have likened the distinctive shape of this planetary to a weight lifter's dumbbell, an hourglass, a bow tie, or even an apple core. Let your imagination run wild." "Easy, large, & rectangular with rounded corners." "Prominent twin lobes & faint halo." "In a modern 6-inch telescope with low or medium powers it appears oval or hazily rectangular with smoothly rounded corners." In 3-inch & larger glasses at 30X to 90X a beautiful, puffy white cloud - very large, bright, & pinched near the middle. Like a celestial pillow floating serenely among the stars of the MW in a 13-inch at 145X! In big amateur instruments there's a "Faint nebulosity which fills the lateral concavities of the body & converts them into protuberances so as to render the general outline of

....
....
....
....
....
....
....
....
....
....
....
....
....
....
....
....
....
....
....
....
....
....
....
....
....
....
....
....
....

the whole nebula a regular ellipse." "View thru 24-inch will leave you searching for your socks!" A "Magnificent & singular object...situated in a crowded vicinity where field after field is very rich...truly one of those splendid enigmas which...are proposed by God but never to be subject to human solution."!? "A beautiful appearance on magnification 94X...very large & shining; two objects blending into one another." "A superb planetary...glowing quite greenish - one of the few planetaries to show vivid (??) color in a small telescope. At low X, when the air is not too steady, the Dumbbell may seem three-dimensional & suspended in space, but this illusion is rare." The rich star field in which it's embedded is "probably responsible for the 'hanging in space' effect." "The most spectacular of its kind in the entire sky. Visible in 7X glasses as a hazy rectangular smudge. Higher magnifications begin to reveal its celebrated hour-glass form & azure color." "Appears as a disk of faint luminous haze, with the interior intensified in the rough form of a cotton-reel or spool...most conspicuous of all PNs." "Remarkable PN, a huge, delicate, crumbled bubble." "Giant 'Dumb-bell', so called on account of its resemblance with low X, but with higher magnification the likeness vanishes, as in the case of most celestial analogies to terrestrial commonplaces. Faint nucleus, many stars enmeshed like pearls caught in swirls of lace. Planetary stage of nova of remote ages." "In a rich field we find two oval hazy masses in contact, of which the preceding (or W portion) seems to me the brighter." "Beautiful, bright & famous." "Like an out-of-focus bow tie, in a rich field of stars. On a good dark night, this ethereal glow of light seems to 'hang in space' among the surrounding stars. One of the more striking & prettier sights in the sky. Take it slowly, & the nebula will reward you." "An ellipse with faintly luminous notches." "Obviously a double con-densation, but bearing little resemblance to its prototype." "Very large & bright, picked up with very low X as two hazy patches of light. Assumes a dumbbell appear-ance in larger (?) apertures, & a complete disc can be photographed." "The Earl of Rosse's 3-foot (36-inch) speculum (metal-mirrored reflector) was thought to reach its starry components: his 6-foot (72-inch) surrounds it with an external ring having a neck like a retort." One of Leland Copeland's 'Major Four' planetaries, along with the Ring, Owl & Crab (actually a SN remnant). "Second largest planetary (in apparent size, after the Helix), greenish color." Actual diameter is about 3LY. "The double-headed shot or dumb-bell nebula...wondrous object." "The central star of the Dumb-bell is very probably a physical double (comp. 17th-mag. at 6"). The primary...is the illuminating star of the nebula; its strong ultraviolet radiation excites the glow of the highly rarefied gases, producing that eerie pale blue fluorescence. The observer who spends a few moments in quiet contemplation of this nebula will be made aware of direct contact with cosmic things; even the radiation reaching us from the celestial depths is of a type unknown on Earth." D=1,200LY The closing words of Webb in his classic work seem appropriate ones on which to end this showpiece roster: "Some of my readers may perhaps feel that I have allotted an undue proportion of space to minute & inconspicuous objects. It may be so. I may have erred in supposing that others might receive as much pleasure as myself from their contemplation: yet a

....
....
....
....

multitude...have been passed by, as well as a great mass of remarks on the beauty or singularity of those which have been selected. But should I have failed in communicating to others a portion of my own interest as to some parts of this list, it will be closed with a nebula (the Dumbbell) which I think will not be found disappointing." Amen!

CONSTELLATION ABBREVIATIONS

(Those in **bold type** are included in this survey, followed by their corresponding page number)

AND **Andromeda - 3**	CRU Crux	**OPH** **Ophiuchus - 46**
ANT **Antlia - 95**	**CYG** **Cygnus - 23, 95**	**ORI** **Orion - 49, 96**
APS Apus	**DEL** **Delphinus - 26**	PAV Pavo
AQR **Aquarius - 5**	DOR Dorado	**PEG** **Pegasus - 55**
AQL **Aquila - 6**	**DRA** **Draco - 26**	**PER** **Perseus - 56**
ARA Ara	**EQU** **Equuleus - 28**	PHE Phoenix
ARI **Aries - 7**	**ERI** **Eridanus - 28**	PIC Pictor
AUR **Auriga - 7**	**FOR** **Fornax - 29, 95**	**PSC** **Pisces - 59**
BOO **Bootes - 9**	**GEM** **Gemini - 29**	**PSA** **Pisces Austrinus - 60**
CAE Caelum	GRU Grus	**PUP** **Puppis - 60**
CAM **Camelopardalis - 10, 95**	**HER** **Hercules - 31, 95**	PYX Pyxis
CNC **Cancer - 11**	HOR Horologium	RET Reticulum
CVN **Canes Venatici - 12**	**HYA** **Hydra - 34**	**SGE** **Sagitta - 62**
CMA **Canis Major - 13, 95**	HYI Hydrus	**SGR** **Sagittarius - 63, 96**
CMI **Canis Minor - 15**	IND Indus	**SCO** **Scorpius - 68**
CAP **Capricornus - 15, 95**	**LAC** **Lacerta - 36**	**SCL** **Sculptor - 73**
CAR Carina	**LEO** **Leo - 36**	**SCT** **Scutum - 74**
CAS **Cassiopeia - 15, 95**	LMI Leo Minor	**SER** **Serpens - 75**
CEN **Centaurus - 17**	**LEP** **Lepus - 38**	**SEX** **Sextans - 77**
CEP **Cepheus - 18, 95**	**LIB** **Libra - 39, 96**	**TAU** **Taurus - 78, 96**
CET **Cetus - 19**	**LUP** **Lupus - 40**	TEL Telescopium
CHA Chamaeleon	**LYN** **Lynx - 40**	**TRI** **Triangulum - 81**
CIR Circinus	**LYR** **Lyra - 41**	TRA Triangulum Australe
COL Columba	MEN Mensa	TUC Tucana
COM **Coma Berenices - 20**	MIC Microscopium	**UMA** **Ursa Major - 82, 96**
CRA **Corona Australis - 22**	**MON** **Monoceros - 44**	**UMI** **Ursa Minor - 86**
CRB **Corona Borealis - 22**	MUS Musca	**VEL** **Vela - 86**
CRV **Corvus - 22**	NOR Norma	**VIR** **Virgo - 87**
CRT **Crater - 35** (under HYA)	OCT Octans	VOL Volans
		VUL **Vulpecula - 92**

ADDENDUM

The following 14 deep-sky wonders (8 of them more neglected and overlooked lovely red stars!) are being added in this expanded edition. Due to the word-processing software that was originally used to write *Celestial Harvest*, it unfortunately was not possible to simply insert them under their respective constellations in this second, revised printing without completely disrupting the paragraph breaks at the top and bottom of all subsequent pages in the manuscript. The author regrets any inconvenience this causes readers.

Object	Coords	Type	Mag	Size	Description
U ANT (1)	10 35 -39 34	SS	5.4-6.8 N (C5)	---	A far-S, red carbon star. Among the brightest of its class, it's seldom looked at due to its low altitude for N Hemisphere observers. Needs a steady, transparent night.
ST CAM (0)	04 51 +68 10	SS	7.0-8.4 N5 (C5)	---	Ruddy carbon star described as "Very deep orange." Looks quite red to most eyes. Visible in a 3-inch glass, best seen in 6-inch or larger scope on dark transparent night.
W CMA (0)	07 08 -11 55	SS	6.4-7.9 N (C6)	---	Another red carbon sun, located NE of Sirius near the CMA-MON border. Provides nice contrast with the many blue-white field stars found in this area of the sky.
RT CAP (1)	20 17 -21 19	SS	6.5-8.1 N3 (C6)	---	Red semiregular variable with approximate period of 393 days. Visible in binoculars throughout its cycle, needs 4-inch or more aperture to appreciate its lovely warm hue.
Struve 3053 CAS (7)	00 03 +66 06	DS	5.9, 7.3 G0, A2	15"	"Almost rivals the famous Albireo." Indeed - a beautiful miniature of Beta CYG & an unexpected surprise! A lovely sight in a 5- or 6-inch at 50X! "Very yellow, blue."
NGC 7023 CEP (2)	21 02 +68 12	DN	6.8	18'	"One of the brightest reflection nebulae" in the sky! "Real bright - I question why it is not plotted in *Norton's Star Atlas*." "Large unusual nebulosity surrounding 7th-mag. star (its mag. is that listed here)." "No trouble seeing the glow even in moderate light pollution" in 6-inch. "Binocular & small-telescope users will find this nebula to be an interesting view." Herschel classified it as a PN? *Sky Cat 2000* gives 5' OC here also?
V460 CYG (0)	21 42 +35 31	SS	5.6-7.0 N1 (C6)	---	A very unusual ruddy carbon star - an unresolved binary system believed to harbor a black hole! Thinking you may actually be looking at the site of one adds to the treat!
NGC 1365 FOR (2)	03 34 -36 08	SG	9.5	10'x6'	Here's a "Large, classic barred spiral" galaxy in the Fornax Cluster - one considered to be the "Most spectacular example of barred spiral structure in the heavens." "Among galaxies, the barred spirals are the most graceful, & of these NGC 1365 is the finest." Easily spotted in even a small glass on a dark night, the bar itself "can be made out protruding to the N & S of the galaxy's central bright core" in an 8-inch scope. Not seen by Sir William Herschel, nor included in Webb's or Smyth's classic works, due to its low elevation above the horizon for these British observers. "Wonderful."
NGC 6229 HER (3)	16 47 +47 32	GC	9.4	4'	Despite its relative dimness & small size (for a globular), a truly fascinating object for medium to large apertures - one mistaken for a PN by its discoverer Sir William Herschel, as well as by Smyth, Webb, Barns & other observers! "PN, faint with 3.7-inch, but beautifully grouped in a triangle with two 6th-mag. stars." "A fine PN, large,

round, & of a lucid pale-blue hue; but its definition & distinctness are encroached upon by the brilliance of the 6th-mag. stars near it." And "Sea-green in starry triangle" (the colors typical of a planetary!). But it's actually a "Bright condensed GC." "Compact & bright, need aperture to resolve." Visible in a 3-inch glass neatly placed with its two stellar sentries mentioned above. An 8- to 10-inch scope at high power is required to turn this clan into a sparkling stellar beehive; in smaller apertures it could easily pass for a PN! NGC 6229 is extremely remote for a member of its class: D=80,000LY

NGC 5897 LIB (6)	15 17 -21 01	GC	8.6?	13'	"The showpiece of Libra." A large, extremely low-surface-brightness, sparce globular. "A universe of remote suns" in a 10-inch. "Large, faint & loose GC, requires large aperture to resolve." "Large but loosely scattered & faint...unspectacular in small instruments." "Faint & pale - so awfully remote an object" in 6-inch glass. "Fine." A "Most unusual object." "Considered by William Herschel to be a gradation between a cluster & a nebula." Some sources give the visual mag. as faint as 10.9 - it certainly doesn't look as bright as listed here! Needs lots of aperture to enjoy. D=40,000LY
BL ORI (0)	06 26 +14 43	SS	6.3-7.0 N3 (C6)	---	Another lovely red carbon star that rivals its neighbor W ORI in both brightness & depth of color. Nice even in a 3-inch at 30X, it's striking in 6-inch & larger glasses.
AQ SGR (1)	19 34 -16 22	SS	6.7-7.7 N3 (C5)	---	Reddish carbon sun. Easy in 4-inch glass. "Red; fine spectrum." Such terse (or often non-existant) descriptions in the literature show how neglected these lovely gems are!
NGC 1514 TAU (2)	04 09 +30 47	PN	10.9/9.4	2'	* Unusual planetary - a bright central star with very faint nebulosity around it. "Some observers consider this object difficult for an 8-inch" but it is visible in a 4- to 5-inch glass, bracketed by two field stars. Although not a showpiece, it's of great historical significance. Sir William Herschel saw it as "A most singular phenomenon! A star of about 8th-magnitude with a faint luminous atmosphere, of circular form, and about 3 minutes in diameter. The star is in the centre, and the atmosphere is so faint and delicate and equal throughout that there can be no surmise of its consisting of stars; nor can there be a doubt of the evident connection between the atmosphere and the star." Thus he recognized for the first time the existence of "a shining fluid of a nature totally unknown to us." Until this observation, all nebulae were thought to be simply unresolved masses of stars. This brilliant deduction, *based soley on the appearance of NGC 1514 at the eyepiece*, showed nebulae to be gaseous long before the spectroscope actually proved it! Displays a "Wonderful, extraordinary aspect in large reflectors."
VY UMA (1)	10 45 +67 25	SS	5.9-6.5 N0 (C5)	---	Ruddy-orange carbon star above the 'Pointer Stars' of the Big Dipper's bowl, visible year round. Its small range in brightness means it always looks pretty much the same!

For the sake of completeness, the author wishes to acknowledge in this addendum section that two of the sky's most famous diffuse nebulae - the large **California Nebula** just N of Xi Persei and the even bigger **North America Nebula** just E of Deneb (Alpha Cygni) - were rejected as showpieces. While striking in photos, their huge angular sizes result in very low surface brightness, making them extremely tough to see visually in telescopes. Interestingly, both *can* be glimpsed with binoculars and even the unaided eye!

SOME THOUGHTS FOR CONTEMPLATION

"The true value of a telescope is how many people can view the heavens through it." - John Dobson

"I became an astronomer not to learn the facts about the sky but to feel its majesty." - David Levy

"The amateur astronomer has access at all times to the original objects of his study; the masterworks of the heavens belong to him as much as to the great observatories of the world. And there is no privilege like that of being allowed to stand in the presence of the original." - Robert Burnham, Jr.

"What we need is a big telescope in every village and hamlet and some bloke there with that fire in his eyes who can show something of the glory the world sails in." - Graham Loftus

"The best thing that we're put here for's to see; the strongest thing that's given us to see with's a telescope. Someone in every town seems to me owes it to the town to keep one." - Robert Frost

"Lastly, remember that the telescope is a scientific instrument. Take good care of it and it will never cease to offer you many hours of keen enjoyment, and a source of pleasure in the contemplation of the beauties of the firmament that will enrich and ennoble your life." - William Tyler Olcott

"The telescope is the lifter of the cosmic veil...a source of far-reaching enlightenment." - Louis Bell

"A telescope is a machine that can change your life." - Richard Berry

"The pleasures of amateur astronomy are deeply personal. The feeling of being alone in the universe on a starlit night, cruising on wings of polished glass, flitting in seconds from a point millions of miles away to one billions of lightyears distant...is euphoric." - Tom Lorenzin

"Of all tools, the observatory (telescope) is most sublime." - Ralph Waldo Emerson

"Adrift in a cosmos whose shores he cannot even imagine, man spends his energies in fighting with his fellow man over issues which a single look through this telescope would show to be utterly inconsequential." - Palomar 200-inch Hale Telescope Dedication

"The appeal of stargazing is both intellectual and aesthetic; it combines the thrill of exploration and discovery, the fun of sight-seeing, and the sheer joy of firsthand acquaintance with incredibly wonderful and beautiful things." - Robert Burnham, Jr.

"Stargazing is that vehicle of the mind which enables us all to roam the universe in what is surely the next best thing to being there." - William Dodson

"The stargazer who takes out his (star) map on a clear night to study the heavens joins a fraternity of the mind chartered at the dawn of time." - National Geographic Society

"Were I to write out one prescription designed to help alleviate at least some of the self-made miseries of mankind, it would read like this: One gentle dose of starlight to be taken each clear night just before retiring." - Leslie Peltier

"What we see through the giant telescopes is an expression of joy by the Creator." - Sherwood Eliott Wirt

"The soul of Man was made to walk the skies." - Edward Young

"Astronomy is good for people's souls." - Deborah Byrd

"To study the skies and the universe we live in is to salute the magnificence of God." - Margaret Woodfin

"The heavens declare the glory of God and the firmament showeth His handiwork." - Psalm 19

"I never behold the stars that I do not feel that I am looking into the face of God." - Abraham Lincoln

"Let us also never forget that astronomy loses half its meaning for the observer who never lets his telescope range across the remote glories of the sky 'with an uncovered head and humble heart.'.... The study of the heavens from a purely aesthetic point of view is scorned in this technological age." - James Muirden

"The serene art of visual observing.... Even if there were no practical application of visual observing, it would always be a sublime way to spend a starry night." - Lee Cain

"I would rather freeze and fight off mosquitoes than play astronomy on a computer." - Ben Funk, Jr.

"The high-tech devices pervading the market are ruining the spirit of the real meaning of recreational astronomy - feeling a close, personal encounter with the universe." - Jorge Cerritos

"Whatever happened to what amateur astronomers really care about - simply enjoying the beauty of the night sky?" - Mark Hladik

"Seeing through a telescope is 50% vision and 50% imagination." - Chet Raymo

"Such is eminently the right use of the telescope...a more extensive knowledge of the works of the Almighty...of the immediate relation between the wonderful and beautiful scenes which are opened to our gaze, and the great Author of their existence." - T. W. Webb

"Never think of the word *amateur* (astronomer) as pejorative. It comes from the Latin *amare*, 'to love,' or, more precisely, from *amator*, 'one who loves.' But then, you knew that all along." - Tom Lorenzin

"But to know is not all, it is only half; to love is the other half." - John Burroughs

"To me, telescope viewing is primarily an aesthetic experience - a private journey in time and space." - Terry Dickinson

"Spend your nights getting intoxicated with photons!" - Telescope Advertisement

"There's something communal and aesthetically rapturous about original archaic photons directly striking the rods and cones in my eyes through lenses and mirrors.... These same photons now impinging on my retina left ancient celestial sights millions of years ago." - Randall Wehler

"Time spent with 2-billion-year-old photons is potent stuff." - Peter Lord

"I was traveling with connoisseurs of blurs, aficionados of the night's subtlest lights. 'Ye littles, lie more close!' That line from a poem by Theodore Roethke seems to me an appropriate prayer for a searcher of the night." - Chet Raymo

"I am because I observe." - Thaddeus Banachiewicz

"Imagination is more important than knowledge." - Albert Einstein.

"Lo, the Star-lords are assembling, And the banquet-board is set; We approach with fear and trembling, But we leave them with regret." - Charles Barns

"As soon as I see a still, dark night developing, my heart starts pounding and I start thinking 'Wow! Another night to get out and search the universe.' The views are so incredibly fantastic!" - Jack Newton

"I am a professional astronomer who deeply loves his subject, is continuously in awe of the beauty of nature...(&) like every astronomer I have ever met, I am evangelistic about my subject." - Frank Bash

"To me, astronomy means learning about the universe by looking at it." - Daniel Weedman

"Nobody sits out in the cold dome any more - we're getting further and further away from the sky all the time. You just sit in the control room and watch television monitors." - Charles Kowal (Palomar/Hale Observatories)

"It is not accident that wherever we point the telescope we see beauty." - R.M. Jones

"I know that I am honored to be witness of so much majesty." - Sara Teasdale

"You have to really study the image you see in the eyepiece to get all the information coming to you. Taking a peek and looking for the next object is like reading just a few words in a great novel." - George Atamian

"Gaze into our Time Machine and see images of the past - images of times before the pyramids were even dreamed of. See back toward the birth of the universe and all times in between." - Telescope Advertisement

"Visit faraway and exotic places from your backyard. This year, take the entire family to the Lagoon nebula. It's easy to reach.... So are all kinds of out-of-the-way spots. Like the Pleiades, Jupiter, Saturn and the Andromeda Galaxy." - Telescope Advertisement

"Some amateur astronomers, it is said, experience the 'rapture of the depths' when observing the Andromeda Galaxy." - Sharon Renzulli

"When you're in the observer's cage of the 200-inch, the telescope turning and the stars going by, it's romantic, beautiful, marvelous." - Jesse Greenstein

"Getting out again to see some old friends in the skies...fills me with renewed enthusiasm. I lose the feeling I sometimes get during the hectic week, and rediscover why I love the stars. Observing all seems so natural, so real, so obvious. How could it possibly be any other way?" - Jerry Spevak

"We have enjoyed knowing the stars. We are among the thousands who have found them old friends, to which we can turn time after time for refreshing thoughts and relief from the worries and troubles of every-day life." - Hubert Bernhard, Dorothy Bennett & Hugh Rice

"The night sky remains the best vehicle of escape I know. Simply...staring up at a crystal clear sky takes the weight of the world off your shoulders." - Victor Carrano

"In desperation, I turn to night as Thoreau turned to his pond...the night is my pond." - Chet Raymo

"The astronomical universe is a beauty to behold, a wonder to contemplate, and a challenge to understand." - Ivan King

"To turn from this increasingly artificial and strangely alien world is to escape from *unreality*. To return to the timeless world of the mountains, the sea, the forest, and the stars is to return to sanity and truth." - Robert Burnham, Jr.

"Consider the sky...overhead on a clear night is the most amazing drama ever offered...the curtain rises...to reveal a superb play written by a divine hand...a wonderful antidote for sanity. We (as astronomers) can always retreat from the turbulence around us to our sanctum sanctorum, the sky." - Max Ehrlich

"A night under the stars...rewards the bug bites, the cloudy nights, the next-day fuzzies, and the thousand other frustrations with priceless moments of sublime beauty." - Richard Berry

"Silently one by one, in the infinite meadows of heaven, blossomed the lovely stars, the forget-me-nots of the angels." - Henry Wadsworth Longfellow

"I celebrate what is up there...I celebrate the most wondrous things.... And there's always that special pleasure in knowing that, when you look upon that distant light, it has traveled all those lightyears - such an incredible journey - just for you. Whenever I contemplate that, I swear it brings tears to my eyes. Yes, I love the universe. Yes, I think it loves me." - Ken Fulton

"How could I convey the mystical love I feel for the universe and my yearning to commune with it?.... The heavens feed us - first the body, then the mind, then the soul.... Gazing into the beginning of everything, we are young once again. The child within us is set free. Remembering the joy of a time we thought would last forever, we play in the starry night." - Ron Evans

"Astronomy is a typically monastic activity: it provides food for meditation and strengthens spirituality." - Paul Couteau

"But it is to be hoped that some zealous lover of this great display of the glory of the Creator will carry out the author's idea, and study the whole visible heavens from what might be termed a picturesque point of view." - T.W. Webb

"Immense have been the preparations for me." - Walt Whitman

"We have loved the stars too fondly to be fearful of the night." - Sarah Williams (epitaph on crypt of John Brashear and his wife)

"If the stars should appear one night in a thousand years, how men would believe and adore and preserve for many generations the remembrance of the City of God which had been shown! But every night come out these envoys of beauty and light the universe with their admonishing smile." - Ralph Waldo Emerson

"A call for reconciliation between science and religion...on the common ground of reverence for the magnificence of the universe...." - Carl Sagan (*Contact*)

"I believe that in looking out at the stars we meet deep psychological and spiritual needs." - Father Otto Rushe Piechowski

"Surely above the starry firmament a loving Father must dwell." - Ludwig van Beethoven

"The stars that we love best are the ones into whose faces we can look for an hour at a time...." - Martha Evans Martin

"All galaxies deserve to be stared at for a full fifteen minutes." - Michael Covington

"The great object of all knowledge is to enlarge and purify the soul, to fill the mind with noble contemplations, and to furnish a refined pleasure." - Edward Everett

"If the pure and elevated pleasure to be derived from the possession and use of a good telescope...were generally known, I am certain that no instrument of science would be more commonly found in the homes of intelligent people." - Garrett Serviss

"The stars bind together all men and all periods of the world's history. As they have seen all from the beginning of time, so shall they see all that will come hereafter." - Alexander von Humboldt

"A star looks down at me, and says: 'Here I and you stand, each in our own degree: what do you mean to do?'." - Thomas Hardy

"A sanctified aura pervades the dome. It seems fitting." - Thomas Mallon (Palomar 200-inch Hale Telescope Visitor)

"Every tint that blooms in the flowers of Summer, flames out in the stars at night." - J.D. Steele

"Why did not someone teach me the constellations in my youth, that I might be at home in the starry heavens, which are always overhead...?" - Thomas Carlyle

"The universe appears to look more like a great *thought* than like a great machine." - Sir James Jeans

"We are dealing with something that resembles a great symphony or a great poem more than it resembles a huge machine." - Robert Burnham, Jr.

"This book is an effort to rescue the ancient love of simple star-gazing from the avalanche of mathematics and physics under which modern astronomy threatens to bury it." - Henry Neely

"But aren't silent worship and contemplation the very essence of stargazing?" - David Levy

"There is something inherently spiritual about looking up at the sky." - Sister Chen

"Oh God, I am thinking Thy thoughts after Thee." - Johannes Kepler

"Celestial observation is a human pursuit almost as natural as breathing, laughing or loving.... The metaphysical questions we address as observers are as profound as those raised when we seek our roots or wonder about the nature of God." - Serge Brunier

"The beloved company of the stars, the Moon, and the Sun...." - Alan Watts

"The hardy Protestants of eastern and middle-western America considered study of the skies akin to godliness, and were eager themselves to look upon the authentic handiwork of so magnificent a Creator." - Edward Pendray

"You are destined to know yourself as a Child of the Living God, heir to the love that created the universe in all its radiant splendor." - Alan Cohen

"I bow to Thee oh God, in the temple of the skies.... Every star of heaven...shall be a window through which to behold Thee." - Paramahansa Yogananda

"A broad and ample road, whose dust is gold and pavement stars, as stars to thee appear seen in the Galaxy, that Milky Way...." - John Milton

"I can never look now at the Milky Way without wondering from which of those banked clouds of stars the emissaries are coming." - Arthur C. Clarke

"In the vastness of space our self-conceit falters. We are humblest when gazing at the stars. We draw nearer to God in contemplation of the immensity of the universe."
- Edgar Cayce

"...the spell with which Astronomy binds its devotees: the fascination and the wonder, not to be put into words, of the contemplation and the understanding of the heavens."
- G. de Vaucouleurs

"The sky...belongs to all of us. It is glorious and it is free." - Deborah Byrd

"To most people the day ends with the setting of the Sun - only the starlovers seem awake to the loveliness and the mystery which surrounds our lives once darkness descends."
- Ben Mayer

"Working peacefully in my telescope shelter as I listened to good music and dreamed about the infinity of the universe." - Hans Vehrenberg

"To gaze into space is to embark upon a spiritual quest, an experience of awe and wonder, a longing for the farthest horizons." - Roger Ressmeyer

"He is glorified not in one, but in countless suns; not in a single Earth, a single world, but in a thousand thousand, I say in an infinity of worlds." - Giordano Bruno

"Astronomy has an almost mystical appeal...we should do astronomy because it is beautiful and because it is fun." - John Bahcall

"Perfect speed, my son, is *being there*!" - Richard Bach (*Jonathan Livingston Seagull*)

"Time is but a stream I go a-fishing in.... I would drink deeper; fish in the sky, whose bottom is pebbly with stars." - Henry David Thoreau

"When I consider Thy heavens, the work of Thy fingers, the Moon and the stars, which Thou hast ordained; What is man, that Thou art mindful of him?" - Psalm 8

"To my Grandson.... When he shall have attained the age of Celestial Inquiry. May he, too, find Star-search In youth a thrilling adventure, In maturity an engrossing problem, In advancing years a compensation and a joy Always!" - Charles Barns

"Spending a dark hour or two working through the starry deeps to catch faint, far trophies is remarkably steadying for the soul. The rest of the world falls away to an extent only realized upon reentering it, coming back with a head full of distant wonders that most people never imagine." - Alan MacRobert

"Stargazing will let you personally tap into the cosmos in a way that will not only relax your tired body and frayed nerves, but at the same time elevate your spirit and let your mind soar as you roam star-struck through the wondrous corridors of creation." - James Mullaney

We stargazers are: "Time Travelers", "Harvesters of Starlight", "Star Pilgrims", "Naturalists of the Night", "Star Hustlers" & "Citizens of Heaven"! - Various Authors

* * * * * * * * * * * * * * * * * * * *

And finally, in tribute to the late Walter Scott (Scotty) Houston - one of the greatest visual observers of all time, and an ever-present source of inspiration to the author and countless other stargazers - here are some thoughts culled from the nearly fifty years of his acclaimed "Deep-Sky Wonders" column in *Sky & Telescope* magazine:

"But let's forget the astrophysics and simply enjoy the spectacle."

"One of the great treasures of life is heaven's starry vault on a clear night, when the familiar constellations blaze forth in mystical splendor."

"How can a person ever forget the scene, the glory of a thousand stars in a thousand hues, the radiant heavens and the peaceful Earth? There is nothing else like it. It may well be beauty in its purest form."

"The beauty of these evenings opens the poetic feelings locked deep inside us all."

"Delightful planetary nebulae - ephemeral spheres that shine in pale hues of blue and green and float amid the golden and pearly star currents of our Galaxy...on the foam of the Milky Way like the balloons of our childhood dreams. If you want to stop the world and get off, the lovely planetaries sail by to welcome you home."

"The celestial actors are in place, a serene majesty washes over the stage, and I can almost hear the music of galactic trumpets in their opening bar."

(The author's rave review of Scotty's collected works, *Deep-Sky Wonders*, appeared in the July, 2000, issue of *Sky & Telescope*.)

* * * * * * * * * * * * * * * * * * *